Construction Contracts

What they said about the first edition:
"A fascinating concept, full of knowledgeable gems put in the most frank of styles . . . A book to sample when the time is right and to come back to when another time is right, maybe again and again."
David A Simmonds, *Building Engineer* magazine

- Is there a difference between inspecting and supervising?
- What does 'time-barred' mean?
- Is the contractor entitled to take possession of a section of the work even though it is the contractor's fault that possession is not practicable?

Construction law can be a minefield. Professionals need answers which are pithy and straightforward, as well as legally rigorous. The two hundred questions in the book are real questions, picked from the thousands of telephone enquiries David Chappell has received as Specialist Adviser to the Royal Institute of British Architects.

The material is considerably updated from the first edition – weeded, extended and almost doubled in coverage. The questions range in content from extensions of time, liquidated damages and loss and/or expense to issues of warranties, bonds, novation, practical completion, defects, valuation, certificates and payment, architects' instructions, adjudication and fees. Brief footnotes and a table of cases are included for those who may wish to investigate further. This will be an invaluable reference for architects, project managers, contractors, QSs, employers and others involved in construction.

David Chappell is Director of David Chappell Consultancy Limited and sometime Professor and Senior Research Fellow in Architectural Practice and Management Research at the Queen's University Belfast and Visiting Professor of Practice Management and Law at the University of Central England in Birmingham. He has written many articles and books for the construction industry, including Spon's *Understanding JCT Standard Building Contracts*, now in its 8th edition. He frequently acts as an adjudicator.

Construction Contracts

Questions and Answers

Second edition

David Chappell

Spon Press
an imprint of Taylor & Francis
LONDON AND NEW YORK

First edition published 2006 by Spon Press

This edition published 2011 by Spon Press
2 Park Square, Milton Park, Abingdon, Oxon OX14 4RN

Simultaneously published in the USA and Canada by Spon Press
270 Madison Avenue, New York, NY 10016, USA

Spon Press is an imprint of the Taylor & Francis Group,
an informa business

Typeset in Sabon & Gillsans by Swales & Willis Ltd, Exeter, Devon
Printed and bound in Great Britain by CPI Antony Rowe,
Chippenham, Wiltshire

British Library Cataloguing in Publication Data
A catalogue record for this book is available from the British Library

Library of Congress Cataloging-in-Publication Data
Chappell, David (David M.)
Construction contracts : questions and answers / David Chappell. — 2nd ed.
p. cm.
1. Construction contracts—Great Britain. I. Title.
KD1641.C479 2010
343.41'07869—dc22
2010003818

ISBN13: 978–0–415–56650–6 (pbk)
ISBN13: 978–0–203–86146–2 (ebk)

Contents

Preface to the second edition

I have been pleased with the reception given to this book since the first edition was published. Since then, I have given further thought to the kind of questions which should be addressed by a book such as this and the result is that the number of questions has been expanded from 125 to 200 arranged in an increased number of sections to make it easier to find all the questions concerned with each topic. I have achieved that by the deletion of 33 questions, which seemed to be least useful, and the addition of a further 108 questions. More than 70 additional legal cases have been included.

The Royal Institute of British Architects' Information Line was set up on 1 May 1995. The idea was that RIBA members could ring in with a problem and be directed to a specialist adviser who would give ten or fifteen minutes of free, liability-free comments to point the architect in (hopefully) the right direction. I have been a specialist adviser to the RIBA since the inception of the service and more recently to the Royal Society of Ulster Architects, answering thousands of questions posed by architects. In my career as a consultant I have also dealt with a multitude of problems from contractors, subcontractors and building owners.

This book includes some of the more common questions, together with a few unusual ones and several misconceptions. Sometimes, there is a ready-made answer, either in the relevant contract or in the judgment of a court. Other questions have no ready answer and in such cases, I have offered a view. Within the limits imposed by the need to keep each answer reasonably short, I have endeavoured to make the answer to each question self-contained. This has resulted in occasional instances where answers have overlapped slightly when dealing with similar subject matters.

Some of the questions were concerned with earlier forms of contract, but they have all been updated as necessary to refer to the latest 2005 series of JCT contracts, i.e. SBC, IC, ICD, MW, MWD and DB and the second revision of the contracts has been considered in each instance. Questions have been included on related topics such as architects' fees, design and disputes. In the four years since the first edition, many important judgments have been handed down by the courts and the process of adjudication has developed considerably.

At the time of writing, the Local Democracy, Economic Development and Construction Act 2009 has received Royal Assent. Part 8 amends the Housing Grants, Construction and Regeneration Act 1996. Among the topics amended are the requirement for construction contracts covered by the Act to be in writing, the notice provisions prior to payment, the adjudication provisions and suspension of performance. No date has yet been fixed for the commencement of the 2009 Act. It is not known whether it will be effective by the date of publication of this book, but it will apply only to contracts entered into after the commencement date. Therefore, in the questions which follow, reference is made to potential changes to the 1996 Act where such changes may be relevant. It is expected that the JCT will issue appropriate revisions to its contracts, other standard forms will be revised and that the Scheme for Construction Contracts (England and Wales) Regulations will be amended.

At the time of writing the RIBA has not published the 2010 editions of its standard appointment documents, but I am informed that publication is imminent.

In writing this book, legal language has been avoided, but reference has been made to legal cases and the relevant citations given so that anyone interested may do some further reading. All these references have been updated. A full table of cases is included at the back of the book.

I appreciate that many building contracts are administered by construction professionals who are not architects. I toyed with the idea of using the term 'architect/contract administrator' or even 'A/CA', but finally decided that for simplicity I would retain the reference to 'architect' when dealing with contract administration. The contractor is assumed to be a corporate body and has therefore been referred to as 'it' throughout.

This book should be useful to architects, project managers, quantity surveyors, contractors, students and those building owners who are anxious to understand more about the workings of building contracts. I have heard a rumour, possibly unfounded, that some lecturers have been using the book as a basis for examination questions.

David Chappell
Wakefield
May 2010

Abbreviations used in the text

ACA 3	Association of Consultant Architects Form of Building Agreement Third Edition
AI	Architect's Instruction
C-10-A	RIBA Concise Agreement 2010
CE/99	RIBA Conditions of Engagement for the appointment of an Architect
CIArb	Chartered Institute of Arbitrators
DB	JCT Design and Build Contract
GC/Works/1	Government Contract for Building and Civil Engineering Major Works
GMP	Guaranteed Maximum Price
IC	JCT Intermediate Building Contract 2009
ICD	JCT Intermediate Building Contract with contractor's design 2009
JCT	Joint Contracts Tribunal
JCT 63	JCT Standard Form of Building Contract 1963 Edition
JCT 80	JCT Standard Form of Building Contract 1980 Edition
JCT 98	JCT Standard Form of Building Contract 1998 Edition
MW	JCT Minor Works Building Contract 2009
MWD	JCT Minor Works Building Contract with contractor's design 2009
NEC	Engineering and Construction Contract
PPC 2000	Project Partnering Contract 2000
RIBA	Royal Institute of British Architects
RICS	Royal Institution of Chartered Surveyors
SBC	JCT Standard Building Contract 2009

SFA/99	RIBA Standard Form of Agreement for the appointment of an Architect
S-10-A	RIBA Standard Agreement 2010
WCD	JCT Standard Form of Building Contract 1998 Edition with Contractor's Design.

Chapter 1

Tendering

1 Can the lowest tenderer legally do anything if its tender is not accepted?

Most invitations to tender contain a proviso that the employer does not guarantee to accept the lowest or any tender. It has long been thought that this allowed the employer considerable freedom to award the contract as desired. To some extent that is correct, but it is not the whole story and employers should take care when tenders are invited that they do not leave themselves open to actions for breach of contract.

In *Blackpool & Fylde Aero Club v Blackpool Borough Council*,[1] the Court of Appeal set out the position when tenders are invited. The position is this. The contractor, by submitting a tender, enters into what can best be described as a little contract with the employer on the basis that, in return for the contractor submitting a tender, the employer will deal with the tender in accordance with the procedure set out in the invitation. At the very least, the contractor is entitled to expect that each properly submitted tender will receive proper consideration. An employer who does not properly consider each tender will be in breach of contract.

Unfortunately, it is not uncommon to find that an employer wishes to see all submitted tenders, even a tender that has been submitted after the closing date and time specified in the invitation. Whatever the architect or quantity surveyor might say, the employer may insist on seeing the tender. On discovering, perhaps, that the late tender is lower than the others, the employer will almost certainly wish to accept it; after all, that is the commercial thing to do.

1 [1990] 3 ALL ER 25.

If this tender is accepted, the employer will be in breach of contract, because the others were invited to tender on the basis that only tenders submitted before the closing date would be considered. The submission of tenders created a succession of contracts, each of which included that term. A contractor who learns that the employer acted in breach of contract would be entitled to claim damages. Such damages would certainly embrace all the contractor's costs in preparing the tender. If all the tenderers discovered the breach (and if one did, it is reasonably safe to assume that they all would), the total damages could be considerable.

There may be other stipulations in the invitation, for example about the course of action to be taken if an error is found in the pricing document. Failure to observe these stipulations will also make the employer liable to any tenderers disadvantaged as a result. Quite apart from legal liability, an employer who indulges in this kind of practice will soon find that no contractor is willing to submit a tender on future projects. In a recent case, an employer, received tenders which were invited on the basis that the procedure would be in accordance with the principles of the Code of Procedure for Single Stage Selective Tendering 1996. In an effort to reduce the price, the employer asked the lowest tenderers to reduce their tenders and, as a result, a tenderer other than the original lowest tenderer became the lowest and that tender was accepted. The original lowest tenderer took action through the courts and it was held that the original lowest tenderer was entitled to recover, not only its costs of tendering but also, the loss of the profit it could have expected if it carried out the project.[2]

However, if the employer strictly observes the rules set out in the invitation, neither the lowest nor any other tenderer has grounds for legal action if a tenderer other than the lowest, or even no tenderer at all, is accepted.

Architects and quantity surveyors who find themselves having to deal with clients who show complete disregard for the tender process must seriously consider whether they can continue to act for such clients. Construction professionals must conduct themselves with complete integrity; this should be an end in itself. In addition, professionals who become associated with doubtful tendering practices

2 *J&A Developments Ltd v Edina Manufacturing Ltd, Armoura Ltd and Others* [2006] NIQB 85.

will get an unenviable reputation among contractors with whom they will have to work in the future.

2 The contractor's tender states that it is open for acceptance for 6 weeks from the date of tender, but the contractor withdraws it after 3 weeks citing a suddenly increased workload. Is the contractor liable to the employer for the additional costs of a replacement contractor?

The answer to this question is to be found in the law of contract.

When a contractor submits a tender, it is an offer to carry out the required work for a certain sum. The employer is free to accept the offer, reject it or to attempt to negotiate. Until the offer is accepted there is no contract. The law is that an offer can be withdrawn at any time before it is accepted and there are some rather awkward rules regarding acceptance by post. Therefore, in normal circumstances the contractor can withdraw the tender before it is accepted and, strictly, no reason need be given. The contractor has no liability for any additional costs suffered by the employer.

The position is different if the employer pays the contractor to keep the tender open. Tenders often state that 'in consideration of a payment of £1 (receipt of which is hereby acknowledged) the contractor agrees to keep the tender open for acceptance for a period of x weeks from the date hereof'. The effect of that is to create a little contract between employer and contractor whereby the consideration is the employer's payment of £1 and the contractor keeping the tender open. Effectively, the employer has bought an option for a few weeks to decide whether or not to accept the contractor's tender. A sum of £1 may not seem much, but the law does not require that adequate consideration is given. It is sufficient if the consideration has some value. In this case, a contractor who withdraws the tender after 3 weeks would be in breach of the little contract and the employer would probably be able to bring an action for damages. The damages would be likely to be the additional costs incurred by the employer in engaging another contractor for the work.

Many employers are not aware that the contractor's offer is also brought to an end if the employer rejects it. An employer cannot reject the offer and subsequently, after undergoing a change of mind, decide to accept it after all. In that situation, what the employer may believe to be an acceptance is actually an offer on the part of the employer to form a contract on the basis of the contractor's original offer. No contract is formed until the employer's offer is unequivocally accepted by the contractor.

3 Does the architect have any particular duty to draw the attention of the contractor to onerous terms or amendments in the contract at the time of tender?

If there are onerous or unusual terms or amendments in the contract, the time to bring them to the attention of the contractor is at tender stage so that the terms or amendments in question can be taken into account in the contractor's tender. If the architect waits until after the contract is executed and the contractor has begun or is about to begin work on site, it will be too late.

The position with regard to onerous terms is fairly straightforward. In general, the contractor will be bound by all the terms of the contract that were notified by the employer at tender stage or, at any rate, before the contract was executed. It is usually sufficient if the contractor is notified by means of the bills of quantities or specification. The part referring to the terms applying in each case is called the 'preliminaries'. It is here that the contractor is informed of the contract to be used and of any changes to the clauses, for example a change in the period for payment from 14 days to 28 days. It is immaterial whether or not the contractor actually reads the terms, so long as the existence of the terms is known.

The point about using standard forms of contract or setting out bills of quantities in accordance with the Standard Method of Measurement 7th Edition, is that contractors know what to expect. They know what the clauses say and they know what will be included in the bills of quantities and where. If the National Building Specification is used, even the wording of the various paragraphs can be reasonably anticipated.

If it is thought desirable to introduce changes to the standard contracts by amending clauses or even introducing new clauses, it will usually be good notice to the contractor if they are put in the usual places. The exception is if the change or additional clause is particularly onerous. What constitutes 'particularly onerous' will be decided ultimately by an arbitrator or judge, or temporarily by an adjudicator. No rules can be laid down about what constitutes onerous. Questions to be asked might include whether it removes important rights from the contractor or introduces significant duties, or whether it gives the employer significant new rights or removes some normal duties. However, the architect and the quantity surveyor must do their combined best to establish before tender stage whether a clause is onerous. If it is decided that it is onerous, steps must be taken in the documents to give proper notice or, to put it in plain words, to bring it to the attention of the contractor. For example, it must not be buried away in the small print. Not only must it be where the contractor would normally expect to find it, it must also be highlighted in some way. Perhaps it should be placed at the beginning of the document or, in extreme cases, be referred to in the covering letter inviting tenders. Fifty years ago, Lord Denning famously said: 'Some clauses which I have seen would need to be printed in red ink on the face of the document with a red hand pointing to it before the notice could be held to be sufficient.'[3]

If generally accepted standard contracts are not used, conditions which are printed on the back of a letter without any reference to them on the front may be held not to be applicable.[4] In one instance, a quotation was sent by fax with conditions on the reverse. The reverse of the page was not transmitted and a court held that the reference on the quotation to conditions on the reverse was not sufficient notice.[5] On the other hand, it is not necessary that the conditions are set out in full in the document, provided that proper notice of them is given.[6] That is the usual situation when terms are simply referred to in bills of quantities. Obviously, onerous terms cannot be referred to in this way unless the contractor is given plenty of opportunity to inspect the actual terms. However, it is always advisable to set out onerous terms in full in the tender document.

The architect's and probably the quantity surveyor's (or the project manager's if there is one) duty is owed to the employer and not to the contractor. It is part of that duty to ensure that the contractor is aware of all the terms, so that the contract is properly binding on both the parties. If, after the contract has been signed or a tender has been accepted, an onerous clause is discovered by the contractor in

3 *J Spurling Ltd v Bradshaw* [1956] 1 WLR 461.

4 *White v Blackmore* [1972] 2 QB 651.

5 *Poseidon Freight Forwarding Co Ltd v Davies Turner Southern Ltd* [1996] 2 Lloyd's Rep 388.

6 *Ocean Chemical Transport Inc v Exnor Craggs (UK) Ltd* [2000] 1 Lloyd's Rep 446.

the depth of the tender documents where a contractor might not easily notice it, the chances are that it will not apply. It is not the slightest use for the architect or quantity surveyor to draw it to the contractor's attention at that stage; it will not be one of the terms in the contract.

4 Is the architect responsible if the tender comes in over budget?

Many architects are concerned about what they should do if tenders come in above the client's budget price. The case of *Stephen Donald Architects Ltd v Christopher King*[7] is instructive. Mr King engaged the architects to deal with the redevelopment of a building into a studio and several flats. Mr King subsequently terminated the architects' services because the cost of the work would exceed the budget. Mr King engaged other architects. As part of the counterclaim, Mr King alleged that the flats were negligently designed, being over-elaborate in the use of materials and space and too costly to proceed. Moreover, it was alleged, the architects did not take proper steps to bring the design within the budget.

The court had some sensible things to say about those allegations. The flats were in an area of local authority housing and it was reasonable for the architects to design some luxurious features to attract buyers into an area that might not seem particularly attractive. Although other architects might have dealt with the brief in a different manner, it could not be said that the architects in this case had produced a design such as no reasonably competent architect would have done.

Once it became apparent that the cost would exceed the budget, the court's view was that the architects should have met with the preferred contractor to try to negotiate a lower price. The architects actually did so and succeeded in making a reduction of some £470,000, producing an achievable construction cost of some £1.3 million. At that stage, the court considered that the architects quite properly waited the outcome of the second tender stage. When that was disappointing, they should have approached the other contractors to see if a better price could be obtained. Although the architects started to do that promptly, Mr King engaged other architects at that

7 (2003) 94 Con LR 1.

point. Therefore, there was no substance in the allegations of negligence against the architects.

So many clients become upset when the tender comes in over budget that it is important for architects to make sure that they work within the budget by having regard to the quantity surveyor's estimates of cost. Where clients add to the brief, architects should confirm the additions in writing and advise clients of the fact that costs are increasing. Invariably at tender stage, clients remember only the original cost estimate: they do not remember asking for changes and richer materials.

Obviously, an architect will be responsible if the tenders show a large increase over the client's budget and there has been no interference and no increase in the client's requirements. Obviously, everything must be taken into account. There may be so much work about that it is difficult to get any contractor to tender and all the tenders are considerably above what might normally be expected. On the other hand, an architect should be sufficiently informed, through the quantity surveyor preferably, so that the client can be warned in advance of the likely level of tenders.

Chapter 2

Pre-contract issues

5 The employer is in a hurry to start work. Is there a problem in the issue of a letter of intent?

It is difficult to think of any other cause responsible for more difficulties and disputes in construction contracts than the employer being in a hurry. The employer's professional advisers should firmly disabuse the employer of the notion that construction can be put underway (successfully) without proper preparation.

Commonly, the architect will try to overcome the problems of a premature start by issuing a letter of intent. Usually, the contractor will have submitted a tender and it will be referable to a standard form of contract, specification and possibly bills of quantities. The issue of the letter may be due simply to the fact that the employer cannot wait the additional few days necessary for the preparation and execution of a formal contract. If that is the only problem, it can be overcome by a simple letter of acceptance of the contractor's tender rather than a letter of intent. More often, there is something more substantial preventing the issue of an acceptance letter. It may be a delay in obtaining funding for the whole project or perhaps the tender was too high and reduction negotiations are in progress.

The idea of a letter of intent is straightforward. It tells the contractor that the employer is not in a position to enter into a contract for the work, but that work can begin and be carried out in accordance with the drawings and specification and if the employer has to stop the work the contractor will be paid for what has been carried out.

There are several problems associated with so-called letters of intent:

- If it is not carefully drafted, the contractor commencing work may create a binding contract for the whole of the work on the basis of the contractor's tender. Simply by putting the words 'Letter of Intent' in the letter heading does not produce a letter of intent.
- If the letter is properly drafted, either employer or contractor can simply bring the arrangement to an end without notice. This can cause tremendous problems if work has been proceeding under a letter of intent until the work is almost complete. For the contractor to walk away at that stage is very expensive for the employer.
- Although work done under a letter of intent is commonly valued and paid on the same basis as the contract that was envisaged in the tender, there is no golden rule about it. Indeed, the contractor is normally entitled to a *quantum meruit*, which may be valued in several ways.
- Sometimes the letter of intent is so carefully drafted that both parties are bound by it until the work is completed although that was almost certainly not the intention.

Letters of intent are sometimes referred to as unilateral contracts, or 'if' contracts. That is a contract formed on condition: 'If you build this wall, I will pay you £100.' If the wall is built, I am obliged to pay the £100, but there is no contract until the condition is fulfilled.

A letter of intent may constitute a continuing offer: 'If you start this work, we will pay you appropriate remuneration.' Again, there is no obligation on the other party to do the work and, if it is done, there are no express or implied warranties as to its quality.[8]

Hall & Tawse South Ltd v Ivory Gate Ltd[9] is a good example of the problems that can arise when projects are commenced using what one or possibly both parties thinks of as a letter of intent.

Ivory Gate engaged Hall & Tawse to carry out refurbishment and redevelopment works. It was intended that the contract should be in JCT 80 form with Contractor's Designed Portion Supplement and heavily amended clause 19. The tender provided for two stages. In view of the need to start work on site as soon as possible, Ivory Gate sent a letter of intent to Hall & Tawse agreeing to pay 'all reasonable

8 *British Steel Corporation v Cleveland Bridge Company* [1984] 1 All ER 504.
9 (1997) 62 Con LR 117.

costs properly incurred . . . as the result of acting upon this letter up to the date you are notified that you will not be appointed'. The letter proceeded to explain the work required and evinced an intention to enter into a contract in a specified sum.

Agreement was not quickly reached and Ivory Gate sent a further letter of intent. What was envisaged was that work would commence, contract details would be finalised and the signed building contract would be held in escrow (a situation where the effectiveness of the contract is subject to a condition being fulfilled). Unfortunately the contract documents were never completed. The terms of the second letter of intent were quite detailed, expressing the intention to enter into a formal contract but, pending that time, instructing the building contract Works to commence, materials to be ordered and Hall & Tawse to act on instructions issued under the terms of the building contract. Previous letters of intent were superseded and, if the works did not proceed, Hall & Tawse were to be paid all reasonable costs together with a fair allowance for overheads and profits. Two copies of the letter were provided and Hall & Tawse were to sign one copy and return it. They neither signed nor returned the copy.

The project took about 9 months longer than was planned. Liability was disputed and, therefore, the money due to Hall & Tawse was also disputed. At the time of the trial, the work was nearing completion. The judge referred to the second letter of intent as a provisional contract and said that it had been made when Hall & Tawse accepted the offer contained in it by starting work on site. It enabled the contract administrator to issue any instructions provided the instructions would be valid under the terms of the contract. The judge held that no other contract had come into existence to supplant the provisional contract and the method of determining the amounts due to Hall & Tawse was to refer to the bills of quantities that were to have formed part of the contract. The machinery for valuing the work was to be found in the JCT 80 contract. Under the provisional contract, Hall & Tawse were not entitled to stop work at any time, as would have been the case under a normal letter of intent.

There were two such letters issued in this instance: one was a true letter of intent; the other was actually a contract that determined the rights and duties of the parties. Although it was intended to be provisional until a permanent contract could be executed, the absence of a subsequent permanent contract turned the provisional contract into a permanent contract. A straightforward letter of intent would have entitled the contractor to walk off site at any time and, crucially, it would have entitled the contractor to remuneration on a fair commercial rate basis, which might have exceeded the contract rates.

Manchester Cabins Ltd v Metropolitan Borough of Bury[10] concerned a letter of intent that was not a contract. Tenders were invited on the basis of the JCT Standard Form of Building Contract With Contractor's Design. Although Manchester Cabins submitted a tender, it did not include the Contractor's Proposals. Much negotiation took place and Manchester Cabins produced some drawings. Bury sent a fax which stated: 'I am pleased to inform you that the Council has accepted your tender for the above in the sum of £41,034.24, subject to the satisfactory execution of the contract documents which will be forwarded to you in due course.'

Eventually, it was confirmed that the letter was indeed authority to commence the necessary preliminary works 'subject to the satisfactory execution of the contract documents . . .'. Surprisingly, later on the same day that the confirmation was sent, Bury wrote to Manchester Cabins suspending work, later stating the Council's intention to withdraw from the contract. The court held that there was no concluded contract, because the phrase: 'subject to the satisfactory execution of the contract documents' was included. Although the phrase did not always prevent a contract from coming into effect, in view of the surrounding circumstances it was clear that there was no agreement in this instance.

Starting work on the basis of a letter of intent or terms incorporated by reference are, therefore, clearly recipes for litigation. It is far better for an employer and a contractor to enter at an early stage into a formal agreement in the current JCT or other form accepted by both parties.

6 If a letter of intent is issued with a limit of £20,000, is the employer obliged to pay a higher sum after allowing a contractor to exceed the limit?

Letters of intent commonly stipulate a maximum figure that the employer is prepared to pay. That is perfectly understandable. The employer needs to know the extent of any financial liability. Thus,

10 (1997) unreported

the proposed Contract Sum may be several million pounds, but, pending final agreement on contract terms and other matters, the employer might issue a letter of intent indicating that the contractor may proceed up to a total of £50,000 or whatever sum is deemed appropriate. The idea is that, before the contractor completes work to that value, either the contract is agreed and executed or the work is stopped.

The problem is that, once a letter of intent is issued, both parties tend to forget what it says and simply get on with the project as though a contract had been signed. Then something happens that concentrates minds and there is a dispute. *Mowlem plc v Stena Line Ports Ltd*[11] is a case in point. The letter of intent concept was taken rather far by the issue of some 14 such letters during the course of the Works. Fortunately, the parties agreed that each letter superseded the previous one, otherwise the dispute might have been labyrinthine in its complexity. When Mowlem commenced the carrying out of the work described in each letter, a small contract was formed by which Stena agreed to pay Mowlem a reasonable sum. In each case, the maximum amount of each sum was stated in the letter.

The last letter sent by Stena stipulated a maximum amount of £10 million and a date for completion. The Works were not finished by the due date and Mowlem's position was that the work carried out was worth more than £10 million. Mowlem maintained that Stena had allowed it to continue the Works even though it was clear that the cost was exceeding the amount in the letter of intent, therefore, Mowlem ought not to be bound by the amount in the letter, which should not have any effect once the sum was exceeded. Stena contended that its professional advice was that the work done did not exceed £10 million.

The court had no hesitation in concluding that Mowlem was entitled to be paid the reasonable amounts it could substantiate under the terms of the letter of intent but such amounts could not exceed £10 million in total. It would not make commercial sense if an agreement to a maximum sum could be set aside simply because the contractor continued to work after the due date or after the limit had been reached.

From this it is to be concluded that letters of intent, like other contractual documents, mean what they say. Usually, if the contractor is

11 [2004] EWHC 2206 (TCC).

working to a letter of intent that specifies, say, £20,000 as the limit, this figure will be exceeded at the contractor's peril. Of course, that is subject to the usual overriding proviso that each set of facts must be considered on its own merits. Where the maximum is low and the eventual sum would be many times that amount, it may be held that a contractor who substantially exceeded the maximum would be entitled to payment on the basis that both parties had clearly agreed to ignore the limit and continue the Works, the failure to issue a revised letter of intent or even a formal contract being an oversight. It is suggested that this would particularly be the case if the contractor had actually received payment above the maximum amount.

7 Can pre-contract minutes form a binding contract?

This question was considered in a recent case.[12] The case concerned whether a contract had been entered into between a contractor and sub-contractor. As is common practice, the contractor had arranged a meeting with the sub-contractor to go through a pro-forma agenda and agree all the basics. The meeting duly took place and the contractor produced a set of minutes which confirmed what had been agreed. Prior to the meeting, the sub-contractor had already submitted a price and priced part of the bill of quantities. The contractor issued a letter instructing the sub-contractor to proceed. The letter said that a formal sub-contract would be prepared. The sub-contractor neither signed the letter nor did it sign the contractor's official order which followed some days later.

The court's view was that there was a contract on terms agreed at the meeting and confirmed in the minutes, because all the essential terms of the contract had been agreed between the parties at that time. However, neither party had framed its arguments in that way and, therefore, the court was unable to make that decision. The court, therefore, decided that the agreed terms constituted an offer which was accepted by the contractor when it sent its letter instructing commencement and that the reference to the preparation of a formal sub-contract made no difference.

12 *Cubitt Building and Interiors Ltd v Richardson Roofing (Industrial) Ltd* [2008] EWHC 1020 (TCC).

Whether an agreement is in writing currently is an important question if a contract is to be brought within the Housing Grants, Construction and Regeneration Act 1996. Section 107 of the Act makes clear that it applies only to agreements in writing. Part 8 of the Local Democracy, Economic Development and Construction Act 2009 will remove that requirement, but until the 2009 Act commences, section 107 is crucial. At first sight, it appears that section 107 is drafted very broadly so as to include virtually every contract, even if it could only tenuously be said that it was in writing. For example, a contract made orally, but which refers to a contract in writing satisfies the Act. However, it is now established that it is not sufficient if the contract was made by a mixture of exchanges of correspondence, notes and oral agreements: a complete record of the parties' agreement must be in writing.[13] That presumably means that if the agreement was fairly basic, it is only what was agreed that need be in writing, subject always to the requirement for the essential terms to be agreed. So, for example, the agreement will not be excluded from the operation of the Act if the written agreement makes no reference to an extension of time clause if the parties did not agree such a clause.

There seems to be little doubt that if minutes were prepared of a meeting during which the parties agreed essential terms, and if such terms were all recorded in the minutes, it would constitute a written agreement for the purposes of the Act. If the contractor submitted a written tender and the order to commence work was recorded in the minutes of a meeting, that would also fall within the Act as written evidence of acceptance of a written tender.[14]

8 What date should be put on a building contract?

The date to be put on a building contract is the date the contract is executed. The contract is executed when the last person who must sign has signed.

In practice, building contracts are usually prepared by theemployer's professional advisers, normally by either the architect or the quantity surveyor. Ideally, it should be prepared by the architect, because the architect is the person who is going to have to administer it.

13 *RJT Consulting Engineers v DM Engineering* [2002] BLR 217.

14 *Connex South Eastern Ltd v MJ Building Services Group plc* (2004) 95 Con LR 43.

Clearly, it is always advisable for the contract to be executed before the contractor commences on site. The consequences of failure to enter into a binding contract have been explored in question 5. The question often arises whether a contract that is not executed until halfway through the construction of a project applies only to the part of the project which is constructed after the date of the signature. This sometimes causes the employer to have the contract backdated to a date no later than the date of possession. This is bad practice for several reasons:

- It is stating something that is false – something to be avoided in any circumstances.
- Circumstances may well arise later when it will be important to know the date on which the parties executed the contract. For example, the employer or the contractor may subsequently bring arguments about the date for possession and it may be useful for the other party to be able to show that the contract was signed by both parties many months after possession and that any difficulties could have been raised at that point – indeed, that one or other party could have refused to sign.
- It is unnecessary, because a building contract will be retrospective to cover the whole of its subject matter.[15] Essentially, that is because the parties, even if signing after part of the contract Works have been completed, are signing an agreement to pay for and to carry out and complete the whole of the Works. Therefore, the contract must apply to what has been done as well as what is yet to be done. There may be a sting in the tail here if, for example, the contractor has failed to comply with the requirements of the contract in some of the work already carried out, perhaps because the contractor at that time harboured the idea that the contract would never be signed and, when signing subsequently, forgot the earlier non-compliance.

9 Can there be two employers on one contract?

It is presumed that 'two employers' means two entirely separate persons or companies. For example, two friends may jointly buy an old barn and intend to convert it into two dwellings for their families. They may wish to, jointly, engage the architect and, jointly, enter

15 *Tameside MBC v Barlow Securities Group Services Ltd* [2001] BLR 113; *Atlas Ceiling & Partition Company Ltd v Crowngate Estates* (2002) 18 Const LJ 49.

into a contract with the contractor. They may reason, with some justification, that if architect and contractor can be sure of doing both dwellings, there may be financial economies of scale.

The straight answer to the question is 'Yes'. However, there are considerable difficulties involved. The architect will require instructions from the clients and the contractor will require instructions from the architect during the progress of the Works. Therefore, the clients must agree about everything. Who will be responsible for paying? Someone must actually sign the cheques. Will a joint account be set up and will both clients have to sign each cheque? What if the clients disagree? To whom will the architect send fee accounts? Will both clients' names be inserted in the building contract as 'Employer'. What if one of the clients wishes to spend more money than the other? Must there be separate accounts within both architect's engagement and the building contract? What is the position if the two friends have a spectacular falling out?

Most of these questions suggest awful situations, possibly resulting in nightmarish legal proceedings. The architect might have a dispute with only one of the employers, but be obliged to take action against both, because the 'Employer' in the terms of engagement is identified by two names.

An architect or a contractor who agrees to contract on this basis probably has a death wish – certainly an insolvency wish. Although, like a good many other recipes for disaster, it can be done, it is not a good idea. A far better idea, in fact the only sensible idea from the architect's and contractor's points of view, is for the two persons to enter into a contract between themselves which sets out how they will do everything connected with the project. Most importantly, it will say which of the persons will be entered into the terms of engagement and act as client. It will also say that this same person will be entered into the building contract as the 'Employer'. In that way, both architect and contractor will have one point of contact and, to put it bluntly, one person to sue if things go wrong. How the two persons arrange their own liabilities is their affair, but not something that need concern the architect and the contractor.

10 If the employer wishes to act as foreman, can each trade be engaged on an MW contract?

It is becoming more common for employers to wish to contract separately with the various trades engaged in the construction process.

During the nineteenth century in Britain it was the norm for employers to engage an architect to design a building, following which the architect, on the employer's behalf, would hire the various trades to carry out the work and a site foreman to control it. Subsequently, main contractors became usual, each one engaging sub-contractors or actually employing the relevant tradesmen directly.

An employer who wishes to engage separate trades and act as site foreman must be sure of having the requisite skills. Although the employer no doubt believes that large sums will be saved, which were added to the bill simply by the existence of a main contractor, it is easy to lose those large sums quite quickly if the progress on site develops serious problems.

The fact is that there is no currently available standard contract that precisely caters for this situation. There are, of course, a multitude of sub-contracts but, not surprisingly, they are all designed to be used with a main contract and a main contractor is assumed. The Trade Contract for use with the Construction Management Contract is more likely, although it still assumes the existence of a construction manager.

Many employers in this situation look to the MW contract. Although it is superficially an attractive proposition, there are many pitfalls. The most important point is that it was not written to address the contractual relationship between the employer and each of a number of separate sub-contractors. It assumes that there is but one contractor on the site carrying out the Works. There is an implied term that the contractor has exclusive possession of the site, which would have to be amended to provide that the particular contractor acknowledges that it is simply one amongst several. Otherwise, there will be claims from contractors alleging the others are interfering with the progress of their work. It assumes that instructions will be given by the architect or contract administrator. If the employer really wishes to act as foreman (and has the ability to do so), there must be a clause in the contract to permit the employer to direct each contractor in its work and to require each contractor to comply with such directions. Virtually every clause would require serious amendment. The liquidated damages clause is unlikely to be applicable and certainly there would be problems with the insurance clauses and termination provisions, which cite employer interference as one ground for the contractor to issue a default notice.

In brief, so many amendments would be necessary to the MW contract and so many additional clauses would be required to deal with such matters as attendances and claims for loss and/or expense that it would be easier for the employer in this situation to commission the drafting of a contract especially for the purpose. The employer would then have to face the possibility of trade contractors refusing to contract on unfamiliar terms.

Chapter 3

Possession of the site

11 Can the project manager change the date of possession in the contract?

It is sometimes thought that the architect can issue an instruction to the contractor to change the date of possession in the contract. That view is misguided. The architect can issue only such instructions as are empowered by the terms of the contract (e.g. clause 3.10 of SBC), and changing the date for possession is certainly not empowered by any clause in JCT contracts. With the popularity of the project manager among clients, a slightly different view has taken hold that the project manager can change the date of possession. This is consistent with the view that the project manager, simply on the basis of his or her appointment, has wide powers under the contract. This view is, if anything, even more misguided.

The date for possession is one of the most important terms in the contract. The employer's obligation is to give the contractor possession of the site on that date (clause 2.4 of SBC, IC and ICD). Failure to give such possession is a serious breach of contract unless the employer has exercised the right to defer possession.[16] The status of a project manager is not easy to define; the powers and duties are not obvious, they depend largely on the discipline of the project manager,[17] but also on the terms of appointment. Usually, a project manager is appointed as the employer's representative, rarely as the contract administrator, and it is even rarer for the project manager to be given all the powers of the employer. Effectively, therefore, the project manager, as normally appointed, does not even have

16 *Freeman & Son v Hensler* (1900) 64 JP 260.

17 *Pride Valley Foods Ltd v Hall & Partners* (2000) 16 Const LJ 424.

power to enter site without the permission of the contractor or the authorisation of the architect.

Even if the project manager was appointed agent with full powers by the employer, the project manager would not have the power to unilaterally change the date for possession; the employer does not have that power. Only the employer and the contractor together may vary the terms of the contract.

12 If the employer cannot give possession on the due date, can the matter be resolved by the architect giving an instruction to postpone the Works?

If the employer fails to give possession of the site on the date named in the contract, it is a breach of contract and, if the failure lasts any significant time, it is a breach of a major term of the contract, because without possession the contractor cannot execute the Works. All standard form contracts provide for a specific date for possession. Failure to give possession is a breach not only of the express terms of a contract but also a breach of the term that would be implied at common law in the absence of an express term.

Because default in giving possession is a breach of a major term of the contract, a protracted failure to give possession, and subsequent acceptance by the contractor of the employer's breach, may entitle the contractor to accept the repudiation and to commence an action for damages, which would include the loss of the profit that it would otherwise have earned.[18] Most contractors would balk at taking such drastic action, but they may claim damages at common law for any loss actually incurred.[19]

Architects sometimes try to overcome a failure on the part of the employer to give possession by issuing a postponement instruction under the contract. For example, SBC clause 3.15 allows the architect or contract administrator to issue instructions in regard to the postponement of any work which is to be executed under the contract. However, the power to postpone is not the same as power to defer possession of the site. Although the architect may issue an instruction to postpone all the work on site until further notice, the

18 *Wraight Ltd v PH&T Holdings Ltd* (1968) 13 BLR 26.

19 *London Borough of Hounslow v Twickenham Garden Developments Ltd* (1970) 7 BLR 81.

contractor is still entitled to be in possession of the site from the date of possession stated in the contract. Postponement is when the contractor is on the site and in control of it, but doing no contract work while in control.

The correct procedure, in the event that the employer cannot give possession, is for the employer to formally defer possession (e.g. under clause 2.5 of SBC) if the contract particulars indicate that power to defer applies. The contract envisages only a maximum 6 weeks' deferment period. Obviously, the employer may opt to amend the contract particulars so that a longer period applies, but there will probably be a price to pay in terms of a higher tender price. If the deferment clause is operated, the contractor is entitled to an appropriate extension of time and whatever loss and/or expense may have been suffered. In practice, where an architect attempts to defer possession by issuing a postponement instruction, the contractor will often accept the instruction as deferring possession, because the contract makes provision for extension of time and loss and/or expense following postponement. Moreover, if the suspension of work lasts longer than the period specified in the contract particulars against clause 8.9.2, the contractor is entitled to take steps to terminate its employment under the contract.

13 In a refurbishment contract for 120 houses under SBC. The bills of quantities say that the contractor can take possession of 8 houses at a time, taking possession of another house every time a completed house is handed over. Is the contractor entitled to possession of all 120 houses at once?

The answer to this question proceeds from this: there is an implied term in every building contract that the employer will give possession of the site to the contractor within a reasonable time. This means that the contractor must have possession in sufficient time to enable the completion of the Works to be achieved by the contract date for completion. For example, under the terms of SBC, clause 2.4 stipulates that the contractor must be given possession on the date set out in the contract particulars.

If the employer fails to give possession on the date stated, it is a serious breach of contract. If, as sometimes happens, there is no express term dealing with the topic, a term would be implied and the failure would be a breach of such a term. Failure to give possession is a breach of such a crucial term that if the failure is continued for a substantial period, it may amount to repudiation on the part of the employer. If the contractor accepts such a breach, an action for damages may be started, which would enable the contractor to recover the loss of the profit that would otherwise have been earned. Generally contractors are not anxious to treat the breach as a repudiation, but simply as a breach of contract for which they can claim damages for any loss actually incurred.[20] SBC contains provision in clause 4.23 that allows the contractor to recover such losses through the contract mechanism (clause 4.24.5) and hopefully avoids the difficulties resulting from accepted repudiation. However, it should be noted that the contract provisions do not displace the contractor's right to use common-law remedies if so inclined.

The position envisaged in the question is still quite common, particularly in local authority housing contracts and it was well stated as follows:

> Taken literally the provisions as to the giving of possession must I think mean that unless it is qualified by some other words the obligation of the employer is to give possession of all the houses on 15 October 1973. Having regard to the nature of what was to be done that would not make very good sense, but if that is the plain meaning to be given to the words I must so construe them.[21]

This was a case where the right to possession had been qualified in the appendix to the JCT 63 form of contract. In order to achieve possession in parts under SBC, it is necessary to complete the contract particulars accordingly. Possession as described cannot be achieved by anything in the bills of quantities, because clause 1.3 makes clear that nothing in the bills can override or modify what is in the printed contract. This is a clause peculiar to JCT contracts that still catches out the unwary.

Part of the judgment in *London Borough of Hounslow v Twickenham Garden Developments Ltd*[22] has helped to give rise to

20 *London Borough of Hounslow v Twickenham Garden Developments Ltd* (1970) 7 BLR 81.

21 *Whittal Builders v Chester-Le-Street District Council* (1985) 11 Con LR 40 (the first case).

22 (1970) 7 BLR 81.

the myth that a contractor can be given possession of the site in parts. The court referred to possession, occupation and use as necessary to allow the contractor to carry out the contract. Because this was not something that the court had to decide, the statement does not have binding force.

The idea that a contractor is entitled only to 'sufficient possession' and that, therefore, the employer need give only that degree of possession that is necessary to enable the contractor to carry out work is misconceived. In any event, the contractor is entitled to plan the carrying out of the whole of the Works in any way it pleases.

Although it is not binding authority, the commentary in one of the Building Law Reports sets out the position:

> English standard forms of contract, such as the JCT Form, proceed apparently on the basis that the obligation to give possession of the site is fundamental in the sense that the contractor is to have exclusive possession of the site. It appears that this is the reason why specific provision is made in the JCT Form for the employer to be entitled to bring others on the site to work concurrently with the contractor for otherwise to do so would be a breach of the contract . . .

This is an eminently sensible view. Although an earlier JCT form of contract was under consideration, the view is equally valid in the context of SBC. Whether or not the contractor has been given sufficient possession is a matter of fact. In another case, under the JCT 63 form, although the employers were contractually obliged to give the contractor possession of the site, they could not do so. This was due to a man, a woman and their dog occupying the north-east corner of the site by squatting in an old motor car with various packing cases attached and the whole thing protected by a stockade occupying part of the site. Although the precise period was in dispute, it seems to have been about 19 days before the site was cleared and the contractor could actually get possession of the whole site. The court held that the employers were obviously in breach of the obligation to give possession on the contractual date. The contractor could enter on to the site, but it was unable to remove the rubbish and its occupants and, therefore, the breach was the cause of significant disruption to the contractor's programme.

In another case, it was held that the phrase 'possession of the site' meant possession of the whole site and that, in giving possession in parts, the employer was in breach of contract and the contractor was entitled to damages.[23]

The item in the bills of quantities cannot override what is in the printed contract (clause 1.3) and the contractor is correct in requesting possession of all the houses on the date of possession, because that is what the possession clause (2.4) states.

23 *Whittal Builders v Chester-Le-Street District Council* (1987) 40 BLR 82 (the second case).

Chapter 4

General contractual matters

14 If the employer is in a partnering arrangement with a contractor, does that mean that the SBC does not count?

Partnering is much misunderstood. First, it must be distinguished from 'partnership', which is a legal relationship between two or more persons acting together with a view to profit. Partnership is governed by the Partnership Act of 1890. Partnering is not a legal relationship at all. It is simply the name that some people have given to the way in which a building may be procured. It is not a procurement system itself, because the procurement of the building might be by means of a traditional arrangement or any combination of project management, design and build or management contracting even though partnering is being practised.

Because partnering is very much a recently invented process, it is not susceptible to clear definition. It may comprise one or more of a whole host of constituent parts. For example, the parties may agree an open book policy whereby the contractor agrees to allow the employer's professional advisers access to its books of account so that they can see whether the contractor is making a profit and how much, and the actual cost to the contractor of labour and materials. There is usually a requirement that all parties will act in a spirit of mutual co-operation and trust and will give each other early warning of problems. Sometimes the arrangement includes provision for both parties to share in any savings and sometimes for the employer to shoulder part of a contractor's loss – up to a certain figure. A guaranteed maximum price (GMP) is often a feature. Like many other things, this is usually not as good as it sounds; a GMP is not necessarily the maximum price and is rarely guaranteed. It has been said, with some truth, that partnering is an attempt to go back to the attitudes and values that existed in the construction industry fifty years ago. The problem is that much has changed in fifty years.

Parties who decide to go forward on a partnering basis do so in one of two ways. First, it is possible for them to enter into a partnering contract, that is to say, a contract that contains within its terms various clauses binding the parties to the partnering ethos. Such contracts as the Engineering and Construction Contract (NEC) and the PPC2000 are examples. A difficulty with both these contracts is that they bear little resemblance to the widely used JCT series of contracts or even the less widely used, but excellent, ACA 3 contract. Therefore, these partnering contracts take a great deal of time to assimilate and use properly. They are both very complex and some may say eccentric. But the parties are legally bound by the partnering principles. Another difficulty is that there is little experience of parties being bound by a requirement, for example, to work in a spirit of mutual trust and harmony. If one party believes another is not being friendly, can the injured party bring legal action for breach of contract? It seems unlikely.

Second, parties may achieve partnering by entering into a legally binding contract in the usual way and incorporating anything that they believe should be legally binding, such as price, access to books, sharing savings and so on. At the same time and usually after all parties have spent a day engaged in getting to know one another, they enter into what is commonly called a 'partnering charter'. This is usually signed at the end of the 'getting together' day whereby all parties (including the professionals) enter into nonbinding undertakings that they will work together in a co-operative way. The charter is non-binding, because the kinds of things it contains are precisely those that are difficult to enforce and depend upon all parties understanding that working together in this way is a good way of achieving a satisfactory outcome for all concerned.

The question posed is based upon a complete misunderstanding of partnering. If the parties enter into a contract such as NEC or PPC2000, it will be binding and breaches of the contract will result in one of the parties being able to bring proceedings in adjudication or either arbitration or litigation. If the parties enter into a JCT contract plus a non-binding charter, they will be bound by the JCT contract. However, what the parties have informally agreed to do in the charter may be taken into account when the way in which the parties have carried out their obligations is considered by a judge or arbitrator.

It has been known for some ill-advised parties to attempt to procure the construction of a building without a legally binding contract at all, but simply relying on expressions of goodwill in a non-binding partnering charter. This is somewhat equivalent to an old fashioned shaking of hands, but in today's climate it is the height of folly. In the event of a dispute, there is no written contract, therefore the provisions of Part II of the Housing Grants, Construction and Regeneration Act 1996 will not apply. Part 8 of the Local Democracy, Economic Development and Construction Act 2009 will remove the requirement for construction contracts to be in writing, but until the 2009 Act commences, a written contract is essential. Whether there is indeed a contract of some sort, perhaps formed by oral exchanges between the parties, will be a matter for the courts. Even if binding, expressions of goodwill are not in themselves sufficient to form the basis of a contract. There must be a known price for the work, or a way of calculating it, and there must be an agreement on the work to be done and the period of time in which it must be done. The courts may be prepared to imply some terms into a contract, but they are not prepared to create a contract where none existed.

The age of partnering (however long or brief it is before the next answer to the construction industry's difficulties arrives) is not a time for throwing away binding contracts. Without a binding contract, neither party is obliged to start, let alone finish, anything.

15 What if no one notices the contractor's serious financial error until the contract is executed?

Tendering is a process conducted at breakneck speed, or so it seems to the contractors concerned. The period allowed for tendering, even if it complies with industry accepted norms, is never really long enough for all the complicated obtaining of sub-contract tenders and the taking into account of the myriad of clauses in the specifications, drawings or bills of quantities. It is little wonder that there are mistakes in tenders. The wonder is that there are not more mistakes.

If there is a mistake that is noticed before a tender is accepted, it must be dealt with in accordance with the particular tendering process adopted. For example, if the quantity surveyor has stipulated that tendering will be in accordance with the CIB Code of Practice for the Selection of Main Contractors (1997), the contractor's act in submitting a tender will create a contract whereby the employer, through the quantity surveyor, agrees to properly consider and apply the Code to any tender properly submitted.[24] Failure to apply the Code in these circumstances will entitle the disadvantaged contractors to take action against the employer for damages. What such damages might be in any particular circumstance is open to debate, but it might, at the very least, allow the contractors to claim the cost of tendering.

The situation for the contractor is grim if the employer has properly complied with all the relevant procedures but the contractor's serious mistake has not been noticed until after the contract documents have been executed. In those circumstances, the contractor has no remedy. The situation was considered as long ago as 1927 in *W Higgins Ltd v Northampton Corporation*,[25] when the court sympathised with the contractor, especially because the mistake was 'really brought about by the carelessness of some official of [Northampton] in drawing up the original bill of quantities'. Unfortunately for the contractor, that carelessness did not alter the fact that the contractor had put in a price substantially lower than intended, it had been accepted and the contractor was bound by that mistake.

The situation is different where the employer realises that the contractor has made a mistake in the tender, but tries to conclude a contract embodying the mistake. Such a situation arose in *McMaster University v Wilchar Construction Ltd*,[26] where the contractor inadvertently omitted an important page of the tender. The page included a price fluctuations clause. The contractor soon realised the error and notified the employer which, nonetheless, tried to accept the tender. The court severely criticised the employer and held that the contract was voidable for a fundamental mistake in its formation. The position is probably the same even if the contractor has not given prior notice to the employer if it can be shown that the employer knew of the mistake in any event.

Therefore, contractors must take great care in preparing and submitting tenders. The *McMaster* case is no doubt unusual. In most cases, quantity surveyors or other relevant professionals subject all tenders to detailed scrutiny and pride themselves on finding errors. Nevertheless, contractors should not rely on the expertise of quan-

24 *Blackpool and Fylde Aero Club v Blackpool Borough Council* [1990] 3 All ER 25.
25 [1927] 1 Ch 128.
26 [1971] 22 DLR (3d) 9.

tity surveyors, because if errors are missed and the contract is executed, the erroneous sums become binding on both parties. It is less likely that an error would be in favour of the contractor, unless it was very subtle, or the tender would not be accepted. But if an erroneous price is accepted, it is binding even if the contractor receives an unexpected windfall.

16 What does 'time-barred' mean?

If something is time-barred, it means that the time prescribed for carrying out a particular action has expired. For example, it is possible to have a time bar clause which states that no extension of time will be given unless a notice of delay is submitted by the contractor within 14 days of the delaying event. Therefore, if no notice has been submitted within the 14 days, the contractor's claim for an extension of time is said to be time-barred. Of course, that is just an example and there may be other factors to be taken into account when considering the entitlement to an extension of time.

Usually, when reference is made to time-barred, the reference is to the effect of the Limitation Act 1980. The Act sets out the time limits for the bringing of various kinds of action. In brief, the most important periods in construction work are:

- For actions in simple contract: 6 years from the date of the breach of contract.
- For actions in deed: 12 years from the date of the breach.
- For actions in tort such as negligence: 6 years from the date when the cause of action accrues (e.g. when physical damage occurs).
- For latent damage: either 6 years from the date on which the damage occurred or 3 years from which the party wishing to claim first knew about the important facts. An action cannot be brought after 15 years after the date of the negligent act. This is usually called the 'long stop' and it applies whether or not the important facts were known within the 15 years, and whether or not damage occurs.

It is important to understand that expiry of the periods in the Limitation Act 1980 does not prevent a party suing. But once the relevant period has expired, it is for the party being sued to raise the expiry of the limitation period as a complete defence. Therefore, if a defendant for some reason omits to raise such a defence and pays up after the limitation period has expired, the payment is valid. It should be noted that, in cases of fraudulent concealment (i.e. deliberate concealment of defects), the limitation period does not begin to run until the fraud is discovered or could have been discovered with reasonable diligence.[27]

17 If a contractor does not have a proper contract, but it has carried out work for the same company before on a written contract, will the terms of that written contract apply again?

The short answer to that question is 'No'. This is a common question and a frequent misunderstanding. People who ask this question are probably thinking of something called 'a course of dealing'. As the name implies, it does not involve simply one previous occasion on which the parties have contracted together. Where parties have dealt with each other using the same terms in a substantial number of previous contracts of a particular type and they enter into another agreement about the same kind of things but omit to include reference to the terms, the previously agreed terms may be incorporated into their latest agreement.[28]

Despite the commonly held view, the situation is quite uncommon and the courts will not hold that there is a course of dealing unless all the evidence points to that conclusion. For example, the courts may hold that there is a course of dealing if a contractor has bought supplies of a certain type of brick from the same supplier over the course of a few years and if there have been a dozen similar transactions always using the same terms and conditions, but on the last occasion, both parties failed to complete the necessary paperwork. However, a party is not entitled to rely on a contract used on only two or three previous occasions as governing the parties in a subsequent transaction.

18 The contractor has no written contract with the employer (A). (A) instructed the contractor to do work and asked it to invoice their 'sister

27 1980 Act, s. 32: *William Hill Organisation Ltd v Bernard Sunley & Sons Ltd* (1982) 22 BLR 1.

28 *McCutcheon v David McBrayne Ltd* [1964] 1 WLR 125.

company' (B). The contractor did so and (B) has not paid despite reminders. What remedy does the contractor have?

This issue is in two parts that are not necessarily connected.

It is always a mistake not to have a written contract. Oral contracts depend upon either the parties agreeing on what they said or the presence of reliable witnesses. It should set alarm bells ringing when a person asks for work to be done or services performed, yet is unwilling to enter into a proper contract. The fact that there is no written contract may suggest that there is no contract at all.

There may be what is sometimes referred to as a *quasi-contract*– more accurately a claim in restitution. In this sort of situation, the claim arises when one party gives instructions to another to carry out work or perform services, knowing that the other party carries out that kind of work or performs those services for a charge. In those circumstances, the law may hold that there is an implied promise to pay.[29]

The second and more important part of the question concerns the identity of the paying party. Whether there is a contract entered into orally by the parties or whether we are dealing with a simple promise to pay, there will be a legal obligation that, in simple terms, amounts to the contractor doing work for (A) and (A) having a legal obligation to pay the contractor in return.

Although the contractor would normally expect to invoice (A) for the work, in this instance (A) has introduced a new element by asking the contractor to invoice (B). Despite the fact that (A) has described (B) as a 'sister company', if both are limited companies they are separate legal entities. Therefore, (B) can quite properly refuse to pay, because it has no obligation to pay the contractor under a contract or even a *quasi-contract*. Therefore, the contractor has no grounds for any legal action against (B) to recover the unpaid charges. Unfortunately, legal action against (A) for the unpaid invoices would be equally fruitless, because the invoices were never sent to (A) and (A) has no obligation to pay an invoice addressed and sent to another party.

Sadly, this situation is not at all rare. If the obligation to pay has been properly assigned to (B) with a written agreement to the assign-

29 *Regalian Properties v London Dockland Development Corporation* [1995] 1 All ER 1005.

ment by all parties the situation is different, but assignment in these circumstances is not the norm. A letter from (A) to the contractor directing that invoices should be sent to (B) will not assist the contractor, unless it unequivocally states that a failure to pay on the part of (B) will be made good by (A).

Usually, the contractor's only remedy is to re-invoice (A) with the whole amount owing, wait the prescribed period and then, if not paid, to take legal action for recovery against (A). Obviously, the contractor cannot recover interest on late payment under the Late Payment of Commercial Debts (Interest) Act 1998 for the full period, because the invoice will be the first it has sent to the correct party. A recurring problem concerns whether or not a binding contract has been entered into in any given situation and, if so, in what terms. That is fundamental to the resolution of most problems in the construction industry. Unless the rights and duties of the parties can be pinpointed with reasonable accuracy, it is impossible to say whether they have or have not complied with their obligations.

Stent Foundation Ltd v Tarmac Construction (Contracts) Ltd[30] is an example of this kind of problem. It is also of interest because it concerned the JCT Management Contract.

At the time the problem arose, the management contractor was Wimpey Construction Ltd, later becoming Tarmac. The employer was a firm called Wiggins Waterside Ltd. Stent was the prospective works contractor. Stent was certainly employed to carry out foundation works – the question was: by whom? The reason why the question came before the court was that Stent had a large claim for the cost of dealing with ground conditions, but Wiggins was in receivership.

Under the JCT Management Contract, there is generally no difficulty that the works contractor is in contract with the management contractor under the terms of the JCT Works Contract. In this instance, however, the position was clouded, because tenders had been invited and a letter of intent sent to Stent by the employer before the appointment of Wimpey as management contractor. Wimpey, although not formally appointed, had been involved in the foundation works contractor tendering process and confirmed to Stent that it would be in contract with Wimpey.

Representatives of Wiggins continued sending instructions to Stent. A key document was probably a letter of instruction to Stent

30 (1999) 78 Con LR 188.

stating that they were to start work on the foundations under the letter of intent, but that Stent must enter into a Works Contract with Wimpey. From then on, Stent and Wimpey acted as though a Works Contract had been concluded between them, but Wimpey stated, quite reasonably, that it could not enter into a Works Contract until the formalities of the Management Contract with Wiggins had been completed. In the meantime, Wimpey continued to process applications for payment from Stent as though the Works Contract was in place. In fact, no Works Contract was ever formally executed, despite Wimpey having executed the Management Contract with Wiggins.

The judge held that a binding contract came into existence between Wimpey and Stent when Wimpey entered into the Management Contract with Wiggins. Wimpey and Stent had been agreed about all the important terms of their contract before then and the contract was dependent only on the execution of the Management Contract.

The judge rejected a suggestion by Stent that Wimpey were estopped (prevented) from denying the existence of a contract just because Wimpey had acted in all ways as though a Works Contract had been executed. He said that Wimpey might have acted in this way in the expectation of a contract being agreed, which would then, of course, be retrospective. That is a very important point to remember. It has long been a construction industry myth that if both parties act as though there were a contract, there really will be a contract. It is a truism to say that every case depends upon its own particular facts; there is no doubt that, in considering whether in any particular instance there is a contract, every facet of the conduct of the parties has to be considered. Acting in accordance with a supposed contract's terms may reinforce the conclusion that there is a binding contract, but it will not usually be conclusive without other evidence.

19 Is the contractor bound to stick to its price if it was described as an 'estimate'?

When a contractor submits an estimate, it is likely that it is intended to be an approximate or probable cost. Nevertheless, if the estimate is submitted by the person able and ready to carry out the work, it may very well be treated as a firm offer to do the work which will be converted into a binding contract if accepted by the person to whom it is offered.[31] In those circumstances, the contractor would be obliged to stick to its price.

This situation may be contrasted with an estimate prepared by a quantity surveyor or an architect for the purpose of indicating to a client the likely cost of a project. There is never any question of the estimate being accepted, because neither the quantity surveyor nor the architect is actually saying that they would construct the project for that price. A wise contractor will probably refer to its price as a rough estimate or a very approximate estimate in order to make absolutely sure that the recipient of the offer knows that the figure is not intended to be accurate.

20 If the employer wishes to bring directly engaged contractors on to site to carry out special work, can the contractor refuse admittance?

Once a contractor has entered into a contract to carry out construction Works, there is an implied term that the contractor has sufficient possession of the site to enable the Works to be carried out. In normal circumstances, the contractor would expect to have exclusive possession of the site, certainly where standard form contracts such as the JCT 2005 suite is concerned. Therefore, unless there is an express term in the contract which allows the employer to bring directly engaged contractors on to site, the employer's action would be viewed as an act of gross interference. The contractor has a better right to be on the site than anyone other than the employer. Therefore, it is doubtful that the contractor could actually prevent the employer from bringing other contractors on to site if they were properly authorised by the employer. However, depending on the precise wording of the contract, the contractor would be likely to have a good claim for additional money either as loss and/or expense or possibly damages at common law.

Some JCT contracts make provision for work being carried out by directly engaged contractors. For example, SBC clause 2.7 refers to work to be carried out by the employer or by any of the employer's persons. Delay caused by such work being undertaken could result in

31 *Crowshaw v Pritchard and Renwick* (1899) 16 TLR 45.

the contractor being entitled to an extension of time (clause 2.19.6) and/or loss and expense (clause 4.24.5). There are two situations covered by clause 2.7, the first where the employer has included information within the relevant contract documents which enables the contractor to know what it is proposed that the directly employed contractors are to do and the second, where such information is not included. If the information is included, the contractor must allow the work to be carried out by employer directly engaged contractors. If there is no such information, the work may only be carried out with the contractor's permission.

21 If the employer has paid for materials on site which are subsequently stolen, who is liable?

This question depends on the conditions of contract. SBC clause 2.24 (IC and ICD clause 2.17 is to similar effect) puts the matter very clearly. It provides that if the architect has certified the value of materials on site and the employer has paid for them, the materials become the property of the employer, but the contractor remains responsible for loss or damage to them. However, this provision is made subject to insurance options B or C where they apply. Therefore, if they do not apply, the contractor is responsible for replacing stolen materials at no additional cost to the employer. Where option B (all risks insurance of the Works by the employer) or C (insurance of existing structures and Works in or extensions to such structures) applies, the employer is responsible for insurance of materials on site. Therefore, the claim for the cost of stolen materials is to be made against the employer's insurance.

It will sometimes be argued that the contractor is liable, even if the employer is responsible for the insurance, because there is a clause in the bills of quantities which states that the contractor is responsible for taking all available measures to secure the site. Unfortunately, such a clause will not be effective to overrule clause 2.24 because clause 1.3 provides that nothing in the bills of quantities can override or modify anything in the printed contract form. Therefore, it seems that even if the contractor is in breach of obligations to secure the site so that materials are stolen, the contractor will be liable for the cost of replacement only if it is insuring under option A.

22 Is there a contract under SBC if everyone acts as though there is?

This question has been touched on elsewhere. Many of the problems in the construction industry would be avoided if the parties concerned ensured that there was a proper written contract in place and signed by both parties before work began on site. Anything else is asking for trouble.

Parties to building contracts seem strangely reluctant to settle all the formalities before work begins. Despite all the evidence to the contrary there is a firm belief that once construction starts on site, the formalities of a legally binding contract will not be a problem. Experience shows that this is a misguided view. In answer to queries about the existence of a formally executed contract an architect will often say that the parties did not get around to actually signing the contract, 'but everyone acted as though the contract had been signed'.

In *A Monk Building and Civil Engineering Ltd v Norwich Union Life Insurance Society*,[32] the existence or otherwise of a binding contract was in issue. Essentially, Monk claimed that there was no contract and that they were, therefore, entitled to be paid on a *quantum meruit* basis: Norwich argued that there was a contract and Monk were entitled only to the contractual sum. It was originally intended that the contract would be executed as a deed before work commenced. Draft contracts were scrutinised by Monk, which made various amendments. Eventually, a letter of intent was issued to Monk on behalf of Norwich. Monk made various amendments to the letter, signed and returned it and made arrangements to start work without prejudice to the unresolved contractual position. At the time of starting work, there were many items still unresolved. The project managers wrote to Monk shortly after work commenced with what was later described as an offer, which it was alleged Monk accepted by continuing work.

Discussion continued during the progress of the Works, but without agreement on several terms. But throughout the project, both Monk and the project managers relied on various contract provisions.

The court held that there was no concluded contract between the parties. The project managers had no actual or ostensible authority

32 (1991) CILL 669, upheld by Court of Appeal (1992) 62 BLR 107.

to negotiate a contract on behalf of Norwich, but only authority to issue the letter of intent. Importantly, there was no agreement of all necessary terms. The court said that it was irrelevant that Monk had relied on contractual provisions during the progress of the Works.

Although the reliance by both parties on contract terms might, in some circumstances, indicate that both parties had accepted that they were bound by the contract, that is an assumption which can be overturned by other factors. Key factors are where, as in this instance, there is no evidence of acceptance of the contract and there is clear evidence that important terms remain unagreed.

23 If a contract is described as Guaranteed Maximum Price (GMP), does it mean that is the most the contractor can receive, no matter what changes there are in the project?

Employers often believe that, under a guaranteed maximum price contract, the contractor is giving an absolute guarantee that the maximum price stated in the contract will not be exceeded no matter what the circumstances. This impression is understandable, because the name and sometimes the way in which such contracts are marketed are misleading. Such contracts are almost invariably entered into on a design and build basis, which place most responsibility on the contractor in any event. However, it should be understood that, if the employer is responsible for extra costs, such as variations, the GMP will be adjusted accordingly and inevitably in an upwards direction.

The price under such contracts is neither guaranteed nor maximum. There is no standard GMP form of contract and most are drafted as required. Some contractors have their own forms of contract for this purpose. The intention is to place all the risk with the contractor except for instances where the employer requires variations, or if the employer carries out some act of prevention which results in loss and/or expense. Contractors are sometimes required to take responsibility for all information supplied, including its accuracy. The risk can be significant.[33] Such contracts aim to secure

33 *Mowlem plc (formerly John Mowlem and Co plc) v Newton Street Ltd* [2003] 89 Con LR 153.

greater certainty about the maximum final cost of the project for the employer and they are usually quite effective, but they must not be viewed as putting an absolute limit on the construction cost. The contractor is expected to carry the risks associated with matters such as ground condition, services, weather and changes in legislation.

24 Can a contractor avoid a contract entered into under economic duress?

The straight answer to this question is 'Yes'. The more important question is 'what is economic duress?' because, unless a contractor can recognise it, the question is academic.

In practice, economic duress is not common. The law recognises that there are certain forms of pressure that may be applied to a party which do not amount to a physical threat to a person nor to damage to goods, but which may allow the innocent party to throw off the contract. That kind of pressure will often be applied, when a contract already exists, in order to obtain better terms.[34]

A typical scenario was demonstrated in *Atlas Express Ltd v Kafco (Importers and Distributors) Ltd.*[35] Atlas was a well-known parcel carrier. It entered into a contract with Kafco, a basketware importer, to carry parcels at an agreed rate. The contract, which was recorded by telex, was crucial to Kafco, because it had just entered into a contract to supply basketware to a large number of Woolworth stores. After a few weeks, the carriers' representative decided that the agreed rate was not viable for their company. Atlas knew that Kafco was contracted to Woolworth, which would sue and cease trading with the importers if deliveries were not made. Atlas therefore wrote suggesting that the agreement be updated, and refused to carry any further goods until a fresh agreement was signed.

Kafco protested, but it had no real option but to sign the new agreement. However, when invoices arrived, Kafco paid only the sums calculated in accordance with the original agreement. Atlas took legal action to recover the balance. In deciding in favour of Kafco, the court pointed out that no person could insist on a

34 *Carillion Construction Ltd v Felix (UK) Ltd* (2000) 74 Con LR 144.
35 [1989] 1 All ER 641.

settlement procured by intimidation. Economic duress was recognised in English law and, in this instance, it voided Kafco's consent to the second agreement. In addition, Atlas had not provided any consideration for the revised agreement and consideration was vital for the establishment of a contract or the variation of an existing contract.

A contract is essentially entered into on the basis of the exercise of free will by both parties. It has long been established that physical violence or the threat of it in order to induce a contract will make that contract void or at least voidable at the option of the threatened party. Economic pressure can be as difficult to withstand as physical violence.

Economic duress must not be confused with undue influence, which is founded on a different principle. In cases where there is some kind of confidential relationship, such as between bank manager and client, the court will usually presume undue influence prevented the client from making an independent judgment if there are dealings between the two parties.

It has been held that the following factors (which must be distinguished from the rough and tumble of the pressure of normal commercial bargaining) must be taken into account by a court in determining whether there has been economic duress:

- whether there has been an actual or threatened breach of contract;
- whether or not the person allegedly exerting the pressure has acted in good faith;
- whether the victim has any realistic practical alternative but to submit to the pressure;
- whether the victim protested at the time;
- whether the victim affirmed or sought to rely on the existing contract.[36]

Economic duress is the wielding of economic sanctions to induce a contract. There are many instances in the construction industry where contractors or sub-contractors are persuaded to make savings or carry out additional work in circumstances that border on economic duress.

36 *DSNB Subsea Ltd v Petroleum Geoservices ASA* [2000] BLR 530.

25 In DB, if the employer provides a site investigation report and the ground conditions are found to be different, who pays any extra cost?

Under the provisions of DB, the Employer's Requirements set out the criteria that the contractor must satisfy in preparing the Contractor's Proposals. That the Contractor's Proposals are a response to the Employer's Requirement is made clear by the second recital. The contractor is likely to be able to found a claim against the employer if site conditions are not as assumed in the Employer's Requirements.[37]

If the employer makes a statement of fact, intending that the contractor will act upon it, it is a 'representation'. Such statements are often made in tender documents. Many of the statements made by the employer within the Employer's Requirements will be representations. The contractor will use the information in compiling its tender. Typically, this will include information about the site and ground conditions. If any of the statements of fact are incorrect, they will probably amount to misrepresentations. A misrepresentation is of no importance unless it is made part of the contract or if it was an inducement to the contractor to enter into the contract. In such cases, the contractor may well have a claim for damages at the least. The precise remedies available depend upon whether the misrepresentation is fraudulent, negligent, innocent or under statute.

The remedies used to be restricted to cases of fraud or recklessness but, as a result of the Misrepresentation Act 1967, they now apply to all misrepresentations. The onus lies with the party who made the representation to prove reasonable ground for believing and actual belief, up to the time the contract was made, that the facts represented were true. Section 3 of that Act restricts the employer's power to exclude liability for misrepresentation.

It follows that a contractor may have a claim for misrepresentations about site conditions made during pre-contractual negotiations. It may claim for damages for negligent misrepresentations or breach of warranty or under the Misrepresentation Act 1967 arising out of representations made or warranties given by or on behalf of the employer.[38] However, the efficacy of all claims depends upon cir-

37 *C Bryant & Son Ltd v Birmingham Hospital Saturday Fund* [1938] 1 All ER 503.

38 *Holland Hannen & Cubitts (Northern) Ltd v Welsh Health Technical Services Organisation* (1981) 18 BLR 80.

cumstances. In an Australian case (*Morrison-Knudsen International Co Inc v Commonwealth of Australia*[39]),the contractor claimed that information provided at pre-tender stage 'as to the soil and its contents at the site . . . was false, inaccurate and misleading . . . the clays at the site, contrary to the information, contained large quantities of cobbles'. There was a trial of a preliminary issue and it was concluded that the basic information in the site document appeared to have been the result of a great deal of technical effort on the part of a department of the defendant. It was information that the plaintiffs had neither the time nor the opportunity to obtain for themselves. It might be doubted whether they could have been expected to obtain it by their own efforts in their role as a tenderer. But it was crucial information for the purpose of deciding the work to be carried out.

A representation followed by a warning that the information given may not be accurate will not usually be sufficient to protect the employer, because it is a clear intention to circumvent section 3 of the Act. Indeed, such a statement may convert the representation into a misrepresentation.[40]

A misrepresentation may also amount to a collateral warranty. For example, in *Bacal Construction (Midland) Ltd v Northampton Development Corporation*,[41] which involved a design and build contract, the contractor was instructed to design foundations on the basis that the soil conditions would be as set out in borehole data provided by the employer. The Court of Appeal held that there was a collateral warranty that the ground conditions would be in accordance with the basis on which Bacal had been instructed to design the foundations and that they were held entitled to damages for its breach.

However, care must be taken, because it has been held that a contractor in a design and build situation is not entitled to rely on a ground investigation report that is simply made known to the contractor by a reference to it on a drawing. It was held not to be incorporated into the contract but had been noted simply to identify a source of relevant information for the contractor. In the court's view, that was not sufficient to override a clause in the contract that placed on the contractor the obligation to satisfy itself about the nature of

39 (1972) 13 BLR 114.
40 *Cremdean Properties Ltd v Nash* (1977) 244 EG 547.
41 (1976) 8 BLR 88.

the site and the subsoil. In addition, the judge made the following thought-provoking observation:

> The nature of a ground investigation report is such that it is unlikely, it seems to me, that parties to a contract would wish to incorporate it into a contract between them. All it can show is what was the result of particular soil investigations. If parties did in fact seek to incorporate a ground investigation report into a contract between them, difficulties could arise as to what was the effect in law of so doing. If one has regard to the terms of the [site investigation] report, and in particular to those two parts of the narrative in which there is a reference to ground water, it really is impossible to say that any definite or positive statement of a nature such as could amount to any sort of contractual term was made. Rather, the information given was hedged about with qualifications as to the accuracy and reliability of what was shown by the investigations undertaken.[42]

Quite so. Nevertheless, employers continue to include site investigation reports as part of the contract information given to the contractor before tendering takes place.

26 What are the dangers for employer and contractor in entering into a supplementary agreement?

During the progress of the Works, the employer may decide that further work is required which is so substantial or of such a nature that it is not appropriate to simply issue an architect's instruction. In such instances, the employer's solicitor will sometimes prepare a supplementary agreement for the parties to sign in order to deal with the additional work. Where work is incorporated by means of an architect's instruction, there is no problem, because the work becomes part of the Works and all the terms of the contract which the parties have already executed cover the additional work. Where the work is added by means of a supplementary agreement, it is essential that the agreement has the effect of bringing the additional work within the

42 *Co-operative Insurance Society Ltd v Henry Boot Scotland Ltd* (2002) 84 Con LR 164.

terms of the original contract. If this is not done, the supplementary agreement may be simply a stand alone contract with no relation to the original contract. The effect of that would be that, unless special provisions were written into the contract, the architect could not:

- issue instructions in relation to the additional work,
- vary the additional work,
- certify the value of the additional work,
- deal with defects in such work,
- certify practical completion,
- certify making good of defects,
- issue a final certificate which would cover such work,
- have any powers or duties whatsoever in respect of the additional work.

The supplementary agreement should specify the extent of the additional work, make clear that it is to be considered as part of the Works under the original contract and make any adjustments necessary as a result of adding the work. For example, it will be necessary to adjust the amount of the Contract Sum and to amend the date for completion of the Works. There may well be other minor things to include, but the essential principle to keep in mind is that the additional work should become part of the original Works. Too often, a supplementary agreement which ignores these points leaves the architect in an impossible position, having no powers under the agreement, but with a client expecting contract administration to be carried out in regard to both original Works and additional work in the usual way. It is important that, if a supplementary agreement is contemplated, the employer engages a person who is experienced in construction contracts to draft the agreement. It is not a time for the draftsman to insert as many clever new clauses as possible, but to reflect the original contract.

27 If a contractor must do something 'forthwith', how quickly is that?

It is often believed that instructing a thing to be carried out forthwith means that it is to be done as soon as possible. JCT contracts call for contractors to carry out architect's instructions forthwith and many architects think that the contractor, on receipt of the instruction, must immediately carry it out. In fact the reality is much more mundane, not to say leisurely. From the cases decided by the courts, it seems that most unsatisfactory of situations will apply: the meaning of forthwith will be interpreted differently depending upon the circumstances. The meaning will vary from 'as soon as possible' by conventional methods of business communication if something is to be sent forthwith[43] to 'as soon as is reasonable' if something is instructed to be done.[44]

Therefore, it seems that, under JCT contracts, an architect's instruction to be carried out forthwith is to be carried out as soon as is reasonable in the circumstances. In other words, the contractor is not expected to drop everything in order to comply with the instruction, but to comply as soon as it is reasonable to do so. Obviously, even that is subject to change if the situation warrants it and there may be instances where it is abundantly clear that an instruction is urgent and forthwith means as soon as possible.

28 What is a reasonable time?

The word 'reasonable' is much used in legal documents and often actions are to be done within a reasonable time. Sometimes this is stated quite expressly in the contract, at other times it is left to be implied.

> When the language of a contract does not expressly, or by necessary implication, fix any time for the performance of a contractual obligation, the law implies that it shall be performed in a reasonable time. The rule is of general application . . .[45]

If a period of time is stipulated in a contract for carrying out some action, the law will not overrule the stipulated time even if it can be shown to be unreasonable. What is a reasonable time will depend upon the circumstances of each particular case. It is an expression which tends to be used when it is impossible to specify precisely how much time should be allowed.

It is sometimes argued that if a contract contains clauses setting out specific times for the carrying out of some obligations and

43 *St Andrews Bay Development Ltd v HBG Management Ltd Mrs Janey Milligan* [2003] ScotCS 103.

44 *London Borough of Hillingdon v Cutler* [1967] 2 All ER 361.

45 *Hick v Raymond & Reid* [1893] AC 22.

nothing for others, where no period is stated none is intended and neither a reasonable nor any other specific period of time can be implied. This suggestion is misconceived and a reasonable time will be implied in such circumstances.[46] Clause 2.15 of SBC requires the architect to issue instructions on being given written notice by the contractor of discrepancies. No time period is stipulated, but it is clear that there must be an implied term that such an instruction must be given within a reasonable time otherwise the clause would be pointless.

46 *RM Douglas Construction Ltd v CED Building Services* (1985) 3 Con LR 124.

Chapter 5

Warranties, bonds and novation

29 Can a warranty be effective before it is signed?

There are relatively few cases on warranties and *Northern & Shell plc v John Laing Construction Ltd*[47] settles an important point.

Laing entered into a contract for the construction of an office block. Under the main contract, Laing was obliged to give a warranty in stipulated terms to the company leasing the building from the developer. The successor to this company was Northern. Some years after completion of the building, defects were discovered and Northern relied on the warranty to recover damages from Laing, because the warranty provided that Laing had complied with the terms of the main contract. The warranty stated: '5. This deed shall come into effect on the day following the date of practical completion of the building contract.'

The limitation period for a deed is 12 years and Northern had not started legal proceedings until not long after that period had expired. However, there was a complication in that the warranty had not been signed until 5 months after practical completion and Northern argued that the normal rules applied and the cause of action arose when it was signed. If that was the case, the issue of proceedings would be inside the 12-year period.

The Court of Appeal decided in favour of Laing. It ruled that clause 5 meant what it said. It had been open to the parties to amend the clause when the warranty was signed so as to take account of the fact that the warranty was being signed retrospectively. The fact that the parties did not amend clause 5 indicated that they intended the warranty to come into effect on the day after practical completion.

47 [2003] EWHC Civ 1035(CA).

The lesson to be learned is that, when a contract of any kind is being executed by the parties after the project has begun, it pays to carefully review the wording of the contract just in case any part should be amended. This is frequently the case in regard to dates for possession in building contracts sometimes signed months after possession was given late. In this instance, the effect was that the warranty started at a date before it was signed. If clause 5 had not been included, the warranty would not have been effective until it was signed. Obviously, if it had never been signed, it would never have been effective at all, with or without clause 5.

30 What are 'step in rights'?

Where warranties are provided, particularly to funders of projects, there is often a clause included which gives a particular party the right to 'step in' and replace one of the parties to an agreement between the other two parties. For example an employer may enter into a standard form building contract with a contractor for the construction of a building. A bank may be providing funds to the employer to enable the project to be built. The bank will often require the contractor to execute a warranty in favour of the bank. The essential clause in the warranty will be the step in clause which enables the bank to take the place of the employer in the building contract in certain circumstances. The circumstances are set out in the clause. Usually, it is when particular occurrences entitle the contractor to terminate the building contract or to treat it as repudiated. The contractor is required to notify the bank if it intends to terminate and it must take no further action until the bank has decided whether to step in. The bank only has a specified number of days within which to decide whether to step in.

If the bank does decide to step in, the contractor must accept the bank's instructions as if it was the employer. The contractor is entitled to accept the bank's instructions. The employer is one of the parties to the warranty principally to acknowledge agreement to the step in procedure and to agree that the contractor can treat the bank as the employer. This is to avoid any danger that the contractor will be in breach of the building contract with the employer. It is likely that there will be specific provisions to safeguard the contractor in the event that the employer has not discharged all payments due. Sometimes, the step in procedure is included in warranties given by the architect to a funder or in warranties given by the architect to the employer where the architect is employed by the contractor in a design and build scenario. In the latter instance, the step in rights may be required by the employer in the event that the contractor goes into liquidation and the employer may wish to complete the building using the contractor's architect.

31 JCT contracts do not seem to mention performance bonds. What are they?

Although JCT contracts refer to certain bonds, such as advance payment and off-site materials bonds, they make no mention of performance bonds. The purpose of any bond is to guarantee payment of some compensation, usually in the event that the party on whose behalf the bond is provided fails to carry out some contractual obligation. In the instances noted above they would relate to failures to repay the advance payment in the instalments required or failure to provide the materials or goods for which payment has been certified by the architect and paid by the employer. They are sometimes referred to as guarantee bonds or guarantees. As a guarantee, a bond must be in writing.[48]

A performance bond is a specific kind of bond whose purpose is to provide compensation if the contractor fails to carry out the building contract. Because JCT standard contracts do not include performance bonds or provision for them, an employer requiring such a bond must ensure that the necessary additional clauses are inserted in the contract at tender stage and that a suitable bond is attached to the documents and, in due course, to the contract. A bond is normally executed as a deed by a person referred to as the surety. The surety is usually either a bank or insurance company. It agrees that, if the contractor fails to perform, it will compensate the employer up to an agreed maximum. The maximum amount is usually 10 per cent of the Contract Sum, but it may be more on small contracts. This ensures that the employer will have some money available, for example, if it becomes necessary to engage another contractor to complete the Works, probably at additional cost.

There are two basic types of bond: default bonds, where the employer has to demonstrate that the contractor was in default of its obligations before the surety can be required to pay out, and on

48 Section 4 Statute of Frauds 1677.

demand bonds, where payment can be required from the surety without the employer having to demonstrate any default on the part of the contractor. On demand bonds tend to be favoured by sureties, because they do not have to carry out any expensive investigations or make any judgment about the contractor's performance.[49] A surety is not usually worried about having to pay out, because it will have made sure that it has adequate security from the contractor for such an eventuality The provisions of a bond must be adhered to strictly. Any changes to the terms of the contract between contractor and employer may result in the surety's liability coming to an end.[50] When the surety is asked to pay, it is usually referred to as the employer 'calling in' the bond.

Performance bonds have dates on which they expire, after which the surety's liability to the employer is ended. Commonly, such dates might be practical completion or, increasingly, the date of issue of a certificate of making good defects. Contractors often write to employers asking them to 'release' the bond. What they mean is that they wish the employer to confirm that the surety's liability is ended. This is important to the contractor because, when the surety's liability comes to an end, the contractor's liability to the surety ends also. This can be very important to a contractor who has obtained a surety from its bank. In such circumstances, a bank will often reduce the contractor's potential overdraft by the amount of the bond until the bond is released (it expires).

32 If the architect is novated to a contractor who subsequently goes into liquidation, can the architect be re-novated to the client?

The question is best answered by first considering the nature of novation. Novation commonly occurs in the construction industry where a contract between employer and consultant is replaced by a contract between contractor and consultant on identical terms. It is usually easier, although not quite accurate, to think of novation as removing one party to a contract and replacing it with another third party. Thus, where an architect has been engaged by an employer to carry

49 *Balfour Beatty Civil Engineering v Technical and General Guarantee Co Ltd* [2000] CLC 252.
50 *Holme v Brunskill* (1878) 3 QBD 495.

out architectural services, it may be agreed that the employer will drop out of the engagement and the engagement will be taken over after tender stage by the successful contractor. Novation requires an agreement between all three parties (in the situation just mentioned that would be the employer, the architect and the contractor), but the main difficulties arise because the contractor will not want the same terms in the contract with the architect as the employer. Therefore, to be effective, the novation must make provision for a change in the terms. The benefit of novation is supposed to be that the consultant is made liable for all the design, even for early design carried out directly for the employer and that this liability is owed to the contractor. That may not necessarily be the case, as discussion in the previous question.

The system is supposed to promote a smooth design process, because it simply continues with the same design team involved. In the case of novation, the duty owed by the consultant to the employer is wholly transferred to the contractor. The consultant and the employer owe each other no further duties. The contractor takes on the employer's duties to the consultant. Therefore, the consultant has exactly the same design obligation, but owed to different parties at the two stages. Moreover, there are different obligations to advise during the stages. The employer and sometimes the consultant may forget that in the second stage, the consultant owes no advisory duty to the employer. If the contractor instructs the consultant to change part of the design, the consultant has no option but to comply, because the consultant is now acting for the contractor, even if the consultant knows or believes that the employer does not want that particular change. This is because the contractor has merely sublet the design to the consultant and it is the contractor which has the direct design responsibility to the employer. The consultant has contracted to carry out the contractor's instructions regarding the design. These instructions may be that the consultant must complete the design in accordance with the Employer's Requirements, but they may be simply that the consultant will take the contractor's instructions. Anecdotal evidence suggests that consultants acting first for the employer and then for the contractor encounter considerable difficulties in practice and the best advice to consultants and to employers is to avoid these situations and act for one or the other party exclusively. However the arrangement is managed, the consultant is always placed in a position of possible, and often actual, conflict.

The question recounts a situation where novation has occurred and the consultant is under a contract with the contractor which excludes the employer. The contractor has gone into liquidation. What happens next is determined by the terms of the contract between consultant and contractor. Some contracts provide that, on liquidation, the obligations of both parties come to an end. Some contracts are silent or allow the consultant to terminate at will. If liquidation of the contractor has actually taken place, it will be as though the contractor simply ceased to exist and, therefore, it would be unable to give or withhold consent for the re-novation. However, in this instance, it is assumed that the winding up process has just commenced. The consultant, depending on the terms of engagement, may simply give notice of termination and then enter into a new contract with the employer. Alternatively, the consultant may request the liquidator, on behalf of the contractor, to enter into a novation agreement with the consultant and the employer with the object of replacing the engagement with an engagement between the employer and the consultant. In the latter instance, it is to be expected that the liquidator will be seeking any possible advantage to the contractor and also, possibly, a cash payment.

The difficulties of a novation which seeks to exclude the employer and introduce the contractor instead have already been mentioned. To attempt to reverse the process will require extremely careful drafting of the novation agreement. It is very likely that difficult liability problems will emerge, particularly concerning the period during which the consultant owed duties only to the contractor.

33 In the case of design and build, can the contractor claim from the architect for design errors in work done before novation?

The benefit of novation is supposed to be that the consultant is made liable for all the design, even for early design carried out directly for the employer and that this liability is owed to the contractor. In practice, that may not be precisely the case. It is often forgotten that the contractor is only liable under DB for completing the design, quite irrespective of whether the consultant is liable to the contractor for the whole design. Therefore, the employer's attempt to channel all design responsibility through the contractor will fail unless DB itself is fundamentally amended to make the contractor liable for all the design. An interesting liability situation may arise.

A significant case in this respect was *Blyth & Blyth Ltd v Carillion Construction Ltd*[51] which highlighted some problems which can arise when the parties enter into a novation agreement. Indeed, a certain type of clause in novation agreements is routinely referred to as a *Blyth & Blyth* clause. In this case, the claimants were consultant engineers who were claiming fees. The defendant was a contractor which counterclaimed against the engineers for losses suffered as a result of some alleged breaches of contract. The case is interesting because of the counterclaim. Although the contract between employer and contractor was the JCT 1981 design and build contract, the decisions of the court are equally relevant to the DB 2005 contract. An important clause was inserted into the building contract as follows:

> any mistake, inaccuracy, discrepancy or omission in . . . the design contained in the Employer's Requirements . . . shall be corrected by the Contractor but there shall be no addition to the Contract Sum in respect of such correction or in respect of any instruction of the employer relating to any such mistake, inaccuracy, discrepancy or omission.

Article 2 of the contract was amended so as to place responsibility on the contractor for any design of the Works which had already been carried out. That is important, but not unique. There were several counterclaims, but the court took as an example the claim for additional costs arising from alleged inaccuracies in the information provided as part of the Employer's Requirements and other information provided prior to tendering. It was alleged that the inaccuracies resulted in the contractor having to supply additional materials for which it could not claim additional payment from the employer, because of the amendments to the contract as noted above. The engineer's terms of appointment contained a clause permitting the employer to require the engineers to enter into a novation agreement in a form annexed to the appointment. In due course the novation agreement was executed by the three parties. The novation agreement provided in the usual way that the engineers' liability, whether before or after the novation, would be owed to the contractor just as though the contractor had been named in the appointment instead of

51 (2001) 79 Con LR 142.

the employer. There was a further provision by which the engineers agreed that their services would all be treated as having been performed for the contractor and that the engineers agreed to be liable to the contractor for any breach of the appointment which occurred before the date of the novation.

The court had to decide whether the contractor could claim against the engineers for loss caused to the contractor due to the engineers' alleged breach of their obligation to the employer to provide accurate information for the Employer's Requirements. The court decided that the contractor was not entitled to make such a claim. The court reasoned that the contractor could not claim against the engineers for a breach of duty which occurred before the novation agreement and was concerning a duty owed to the employer, unless the employer had suffered the losses. The novation agreement permitted the contractor to act as though it had been the employer at the time of the pre-novation breaches. In the *Blyth* case an attempt was made by the contractor to give effect to the novation agreement by substituting the word 'contractor' for 'employer' in the appointment whenever it occurred. This produced results which the court described as nonsensical. The reason why the contractor was unsuccessful was mainly because the employer, in whose shoes the contractor attempted to stand, did not suffer any loss. The *Blyth & Blyth* clause, already mentioned, attempts to forestall that defence by requiring that the consultant will not argue in any legal proceedings that a contractor's claim fails simply because the employer did not, in fact, suffer a loss. Some clauses go further and attempt to link directly to losses suffered by the contractor. However, where a normal novation agreement is employed, the failure to include additional clauses on the *Blyth & Blyth* model will probably result in the contractor being unable to claim against the consultant for pre-novation negligence.

34 Have architects any choice whether they provide collateral warranties?

A collateral warranty, sometimes called a 'duty of care agreement' or a 'collateral contract' is a contract with the same legal attributes as other contracts. Foremost is that no one can be obliged to enter into a collateral warranty. Therefore, architects and others from whom warranties are requested have a choice, just as they have a choice whether or not to enter into terms of engagement with a prospective client. These are agreements freely (if at all) entered into. There is no doubt that the provision of a warranty makes the architect liable to more parties than would otherwise be the case.

There are two common circumstance in which architects (and other members of the design team) may be asked to enter into warranties. The first is if the architect has agreed in the terms of engagement to provide a warranty. The second is if the client has asked the architect to provide a warranty although there is no agreement to do so.

If the architect has agreed to provide a warranty, it must be provided. However, the client cannot insist upon a particular form of words in the warranty unless a sample warranty in exactly those words was fixed to the terms of engagement and the architect agreed to execute a warranty in just those words. If the terms of engagement simply refer to a warranty without providing a sample, the architect is in a very strong position and can offer to execute a warranty in any terms at all. In such instances, it may be expected that the architect will have a form of warranty drafted which protects his or her interests. The client is left with a choice to either accept what is offered or to reject it, but the client cannot insist on something different.

If there is no agreement to provide a warranty, a client may still put pressure on the architect to provide one. If the client is a friend, a long standing client or a new client from whom many more commissions are expected, it may prove difficult to resist. Commercial pressures will often overcome contractual rights. However, if the architect does agree to provide a warranty, the position is similar to the situation where the terms of engagement allow for a warranty but fail to specify the wording. In such circumstances, the architect should either insist on using the special protective warranty or make a substantial charge if using the client's preferred wording.

A client's solicitor may try all kinds of arguments to make the architect sign. A favourite one is to say that all the other consultants have signed similarly worded warranties. If true, that is like saying that all the other consultants are happy to enter into onerous warranties and risk future claims so why is the architect not similarly minded. Another argument is that all the clauses are standard clauses. There are no standard clauses in warranties. There are clauses which are repeated in warranties, but most warranties are specially drafted by solicitors for a particular project and lawyers have their own versions of every clause and often add entirely new ones of their own devising. There are one or two standard warranties obtainable, but they are only standard in the sense that someone has drafted the warranty and had it printed for sale.

Even if it was agreed that there are some standard clauses in warranties, it does not mean that such clauses have to be accepted. Every day, lawyers throughout the land are busy amending clauses in standard building contracts because they do not believe that the particular clauses are in their clients' best interests. Some warranties have the initials of various professional bodies included presumably to indicate that they endorse or at any rate do not condemn the warranties concerned. That does not necessarily mean that such warranties will protect the architect's interests. The best kind of warranty for that purpose is no warranty at all.

Chapter 6

Contractor's programme

35 If an architect approves a contractor's programme, can the contractor subsequently change the programme without the architect's knowledge and, if so, can the architect demand an update?

Provision of a master programme by the contractor is covered by SBC clause 2.9.1.2, which requires the contractor to provide two copies to the architect. Approval of the programme by the architect is dealt with in the next question. Suffice to say that it has no particular effect on the contract or contractor's responsibility for the programme.

SBC is the only one of the JCT traditional contracts that actually refers to the contractor's programme. Clause 2.9.1.2 should be carefully studied. It states that the contractor will provide the architect without charge with two copies of the master programme for the Works. The contractor's obligation to revise the programme only occurs if an extension of time is given either by the architect under clause 2.28 in the normal way or by a pre-agreed adjustment as a result of the acceptance of a schedule 2 quotation.

The contractor has no obligation to revise the programme if the contractor falls behind due to its own fault. Looked at stringently, that makes perfect sense. If it is the contractor's fault that a delay has occurred, it is clearly the contractor's responsibility to recover the lost time under clause 2.28.6.1, which requires the use of best endeavours *constantly* to prevent delay. It can be convincingly argued that the programme requires no adjustment because the contractor ought to be doing everything a prudent contractor would do to get back on programme. The only time the programme requires adjustment is if the completion date is adjusted to a later date. In that case, the programme should be amended to reflect completion on that later date.

The practice of contractors constantly submitting revised programmes, not because the completion date has been revised but because the contractor has fallen behind, is to be deplored. The architect should not be interested in when the contractor says it believes it will finish if that date is not the completion date in the contract. The problem is that many contractors take the view that the contractual completion date is simply a date to aim for and that, if it is not achieved, the architect will probably, if sufficiently threatened, extend the completion date to match the date of practical completion. The constant submission of programmes that bear no relation to the contract or extended date for completion does nothing to assist the architect in considering extensions of time or disruptions to the regular progress of the work. For that purpose, a programme submitted at the start of the project that accurately reflects the intended progress and completion is essential.

All the foregoing is by way of putting the question into context. Clause 2.9.1.2 is the only clause that refers to the contractor's programme and it will readily be seen that nothing in the clause states that the contractor must comply with its own programme. First, it should be noted that the reference is to a master programme. That allows the contractor to produce numerous detailed programmes which there is no obligation to provide for the architect. Second, the contractor can opt not to work to its own programme or even opt to change the programme without informing the architect. Only if the change results from an extension of time must the revised programme be submitted. Therefore, it is possible for a contractor to submit a programme indicating that work will progress from point A through B, C, D etc. to Z and, on site, commence at point Z and work in reverse. There is nothing in the contract that obliges the contractor to work to its programme. Most contractors do, of course, but the question was no doubt prompted by the fact that from time to time a contractor will find it useful to significantly vary its work from the submitted programme.

The programme is not a contract document, although it might be termed a contractual document, because it is generated in accordance with a clause in the contract. However, neither the contractor nor the employer is bound to follow it. It would be possible to make the programme a contract document, but that would not necessarily be an advantage, because every slight deviation from the programme potentially would have a financial implication.[52]

Therefore, as the standard contract currently stands, the contractor can change the programme without the architect's knowledge or permission and the architect has no power to require an update. Obviously, additional clauses can be introduced to deal with some of these difficulties and programmes can be required in particular formats. However, considerable thought should be given before deciding to make it a contract obligation for the contractor to comply with its own programme. There could be substantial financial repercussions as noted above.

36 Under SBC, the architect has approved the contractor's programme, which shows completion 2 months before the contract completion date. Must the architect work towards this new date?

It is quite common for a contractor to show a date for completion on the programme that is before the date for completion in the contract. It is never very clear why a contractor should do this, because the programme cannot alter the contract date for completion. All the programme does is to inform the employer and architect that the contractor intends to complete before the completion date. The contractor is entitled to do this because clause 2.4 states that it must complete the Works 'on or before' the completion date.

By asking whether the architect must 'work towards this new date', the questioner can be referring only to three important situations: the most obvious is the provision of further information to the contractor; another is responding to a notice of delay under clause 2.28 where the architect, in deciding whether to give an extension of time, is to consider whether completion of the Works is likely to be delayed beyond the completion date; the final important one is the architect's obligation under the same clause 2.28 to notify the contractor about an extension of time decision no later than the completion date. Other matters relate to the date of practical completion, which ought to be the same as, but which is not necessarily connected to the contract completion date. The contract date for completion is

52 *Yorkshire Water Authority v Sir Robert McAlpine and Son (Northern) Ltd* (1985) 32 BLR 114.

the date by which the contractor undertakes to complete the whole of the Works. The date of practical completion is the date by which virtually the whole of the Works are, as a matter of fact, complete.

To deal with the provision of information first, the courts have held that, although the contractor is entitled to complete before the completion date in the contract, the employer has no obligation to assist the contractor to do so.[53] This principle is now enshrined in the contract at clause 2.12.2. This clause stipulates that, in providing further information, the architect must have regard to the progress of the Works. This means that if the contractor is progressing so as to complete by the completion date, the information must be provided at such times as will enable the contractor to finish the Works on time; but that if the contractor is making slow progress, the architect is entitled to slow the delivery of information to suit. However, the clause proceeds to state that, if the contractor seems likely to finish before the completion date, the architect need only provide information to enable it to complete by the completion date.

The second and third situations deal with extension of time and assume that everyone knows the completion date. The only way in which the architect's responsibilities under clause 2.28 can be affected is if the contractor's programme, showing a completion 2 months earlier than the contract date, somehow takes precedence over the completion date in the contract. If that is the case, it will clearly affect the provision of information also.

The contract date for completion, like the other terms of the contract, can be changed only by agreement between the parties – the employer and the contractor. Approval of the programme by the architect is of little or no consequence. The programme is not a contract document and the parties are not bound by it – not even the contractor, strange as that may seem. Most architects will not approve the contractor's programme, but even if the programme is approved, it merely signifies that the architect is happy with it. In effect the architect is saying: 'I am happy for you to work in accordance with the programme if that is what you wish to do.' Approval does not transfer the responsibility for the thing approved from the contractor to the architect[54] unless the contract makes specific provision for that to happen.

53 *Glenlion Construction Ltd v The Guinness Trust* (1987) 39 BLR 89.
54 *Hampshire County Council v Stanley Hugh Leach Ltd* (1991) 8-CLD-07-12.

The position might be different if the contractor submitted its programme showing completion 2 months earlier than the contractual date and it was discussed and agreed by both parties and the new date confirmed. In that instance, the architect may be obliged to work in every way as though the programme date for completion was the contract date for completion. The confirmed agreement of both parties amounts to a variation of the contract terms. In practice, this could take place at a site meeting; possibly the one usually and incorrectly called the 'pre-contract meeting' ('pre-start meeting' is a better name). Where the parties reach an agreement at such a meeting, with plenty of witnesses and where the agreement is recorded in the minutes which are subsequently agreed by all parties, it is difficult to escape the conclusion that a valid variation of the contract terms has taken place. If the circumstances are such that a valid variation of the terms has not occurred and it is only the architect who has received and commented on the programme, the contractor's obligation is still to complete by the contract date for completion.

As noted earlier, it is difficult to see what the contractor gains by this strategy. If the contractor notifies delay, the architect appears to be able to take the contractor's proposed completion date into account when deciding whether completion of the Works is likely to be delayed beyond the completion date. That would not work to the contractor's advantage. Suppose the contract completion date was 30 September and the contractor's proposed date for completion was 30 July. If the contractor argues that it is being delayed in completion by 2 weeks, there seems to be no reason why the architect should not assume that the contractor will, therefore, complete by the middle of August. That would not indicate a delay to the contract completion date and no extension of time would be due. On this basis, the contractor would have to register a delay of more than 2 months before an extension of time would be due.

37 Can the architect insist that the contractor submits the programme in electronic format?

The rights and obligations of the employer and the contractor are governed by the terms of the contract. None of the standard forms in general use require the contractor to submit a programme in electronic format. Indeed, only one of the JCT series of contracts requires the contractor to submit a programme at all and that is SBC, which requires the contractor to submit a master programme to the architect.

There is no doubt that a programme in electronic format can greatly assist the architect in deciding upon the correct extensions of time, in considering applications for loss and/or expense and in generally monitoring the progress of the Works. Of course, there are a number of different computer programs that will support a building programme and it is essential that any programme submitted by a contractor in electronic format is compatible with the software on the architect's computer. The alternative is to ask the contractor to supply a list of the activities on the programme together with the relevant predecessors and successors so that they can be entered into the architect's computer. This is a rather more tedious operation than simply inserting a disk, but it is far better than having no such programme at all.

All that is necessary to require the contractor to submit an electronic formatted programme is for a suitable clause to be inserted in the preliminaries section of the bill of quantities or specification. This will not conflict with the printed contract in most cases because, as noted earlier, only SBC requires a programme of any kind. It will not conflict with SBC, because the contract does not specify the type of programme, therefore the additional clause is not overriding or modifying the printed form but merely amplifying it by adding further requirements. In this clause, the architect can require the programme in a form to suit his or her own computer software. A contractor who fails to supply the programme in the form specified will be in breach of contract and the architect is entitled to take account of the breach in calculating extensions of time.[55]

The use of computers in dealing with extensions of time has been approved by the courts. What is the position if there is no requirement for the contractor to supply an electronic version of the programme? Can the architect compel the contractor to provide it in any event? The answer to that appears to be that, in the absence of a specific clause in the contract, the architect cannot compel the contractor to provide a programme in any particular form.

However, when a contractor notifies delay in the expectation of receiving an extension of time, there is nothing to prevent the architect asking for a programme in a specific form as part of the architect's general power to request further information from the contractor. The contractor may refuse to provide it, arguing that

55 *London Borough of Merton v Stanley Hugh Leach Ltd* (1985) 32 BLR 51.

there is nothing in the contract that requires it. The architect's reply would be to simply point out to the contractor that, in the absence of such a programme as requested, the architect will find it very difficult to determine a fair and reasonable extension of time and that the contractor will be the author of its own misfortune if it does not receive the extension of time it expects.

That usually has the desired effect. If not, the architect may give a somewhat shorter extension of time than expected and, if this is done as part of the review of extension under SBC, IC or ICD during the 12-week period after practical completion, the contractor will have to live with it, because the architect has no power under these contracts to revisit the situation again once the 12 weeks has expired. Indeed, under MW and MWD, there is no review provision and, therefore, no opportunity for the architect to reconsider after the date for completion in the contract, or as already extended, has passed.

38 When a contractor says that it owns the float, what does that mean?

Float is a commonly used term that is much misunderstood. It is the difference between the period required to perform a task and the period available in which to do it. Critical activities have no float – that is why they are critical: there is no difference between the period needed to carry out the activity and the period allocated for it. In other words, there is no scope for any delay at all before the completion date of the project is affected. To say that an activity has a day of float means that the activity could be extended by another day without affecting the completion date of the project. Put another way, the activity could commence a day late without there being any effect on the overall programme.

Contractors sometimes argue that an extension of time is due even if a non-critical activity is delayed. They argue that they own the float and, therefore, that the employer cannot 'take advantage' of it. This is a very strange contention. No one owns float. It is like trying to argue that a person is taking advantage of the air around them. Float is simply the space before or after individual activities when they are put together in the form of a programme. Whether it actually exists at all depends on whether the programme is accurate.

It is sometimes argued that if a contractor programmes to complete a 12-month contract in 10 months, the 2 months are the contractor's float and, therefore, if the project is delayed by even a week, an extension of time will be due, even if the contractor finishes several weeks before the completion date. That is obviously wrong. If the week's delay causes float to be used, because of some employer default, the contractor has no entitlement to an extension of time because the delay does not affect the completion date; that is the crucial point. It has been suggested that in such circumstances, if the contractor is subsequently delayed for reasons which do not warrant an extension of time and for which it would become liable to pay liquidated damages, the architect should nevertheless give an extension of time for a period not exceeding the length of the float.[56] The point was made by the court *obiter dicta* (words said by the way and not necessary for the decision).

A better view was set out in *Ascon Contracting Ltd v Alfred McAlpine Construction Isle of Man Ltd*,[57] where the judge put the position like this when dealing with a contractor's claim against a sub-contractor:

> . . . I must deal with the point made by McAlpine as to the effect of its main contract 'float' . . . It does not seem to be in dispute that McAlpine's programme contained a 'float' of five weeks in the sense, as I understand it, that had work started on time and had all sub-programmes for sub-contract works and for elements to be carried out by McAlpine's own labour been fulfilled without slippage the main contract would have been completed five weeks early. McAlpine's argument seems to be that it is entitled to the 'benefit' or 'value' of this float and can therefore use it at its option to 'cancel' or reduce delays for which it or other sub-contractors would be responsible in preference to those chargeable to Ascon.
>
> In my judgment that argument is misconceived. The float is certainly of value to the main contractor in the sense that delays of up to that total amount, however caused, can be accommodated without involving him in liability for liquidated damages to the employer or, if he calculates his own prolongation costs from the contractual completion date (as McAlpine has here)

56 *Royal Brompton Hospital NHS Trust v Hammond & Others (No. 8)* (2002) 88 Con LR 1.
57 (2000) 16 Const LJ 316.

> rather than from the earlier date which might have been achieved, in any such costs. He cannot, however, while accepting that benefit as against the employer, claim against the sub-contractor as if it did not exist. That is self-evident if total delays as against sub-programmes do not exceed the float. The main contractor, not having suffered any loss of the above kinds, cannot recover from sub-contractors the hypothetical loss he would have suffered had the float not existed, and that will be so whether the delay is wholly the fault of one sub-contractor, or wholly that of the main contractor himself, or spread in varying degrees between several sub-contractors and the main contractor. No doubt those different situations can be described, in a sense, as ones in which the 'benefit' of the float has accrued to the defaulting party or parties, but no one could suppose that the main contractor has, or should have, any power to alter the result so as to shift that 'benefit'. The issues in any claim against a sub-contractor remain simply breach, loss and causation.
>
> I do not see why that analysis should not still hold good if the constituent delays more than use up the float, so that completion is late. Six sub-contractors, each responsible for a week's delay, will have caused no loss if there is a six weeks' float. They are equally at fault, and equally share in the 'benefit'. If the float is only five weeks, so that completion is a week late, the same principle should operate; they are equally at fault, should equally share in the reduced 'benefit' and therefore equally in responsibility for the one week's loss. The allocation should not be in the gift of the main contractor.
>
> I therefore reject McAlpine's 'float' argument.

This is good authority that float is owned by no one. The decision in *How Engineering Services Ltd v Lindner Ceilings Partitions plc*[58] is to similar effect. Therefore, when a contractor says that it owns the float, certainly under JCT contracts, it means that the contractor does not properly understand the concept of float.

58 17 May 1995 unreported.

Chapter 7

Contract administration

39 Does the contractor have a duty to draw attention to an error on the architect's drawing?

Generally and in normal circumstances, the contractor has no liability for design, therefore no liability for the production of design drawings. The question often arises whether the contractor is entitled simply to build what the drawings and specifications set out, even if there are errors on the architect's drawing. It is not surprising that most architects would say 'No', but the case law on this subject is not so clear.

It has been established by a Canadian case that a contractor will be liable to the employer for building errors in a design if the original architect was not involved in the construction stage.[59] The *ratio* of that case seems to have been that the employer was no longer relying on the architect and, therefore, relied solely on the contractor, which should have taken care to check that everything on the original architect's drawing worked properly. That, however, is not the situation under consideration here, where the original architect is still engaged, but where there is an error in the drawings. There were two cases in 1984[60] which held that a contractor did have a duty to warn the architect if it believed that there was a serious defect in the design. Subsequently, however, another court decided that such duty as the contractor might have was to the employer and probably only in those cases where the contractor was aware of the employer's

59 *Brunswick Construction v Nolan* (1975) 21 BLR 27.

60 *Equitable Debenture Assets Corporation Ltd v William Moss* (1984) 2 Con LR 1; *Victoria University of Manchester v Wilson and Womersley and Pochin (Contractors) Ltd* (1984) 2 Con LR 43.

reliance for at least part of the design.[61] This has echoes of the Canadian case mentioned above. To further confuse matters, another case held that a contractor had a duty to at least raise doubts with the architect if there appeared to be something wrong with the drawings.[62] One would have to wonder at the motives of a contractor who had full knowledge of a drawing error and yet failed to draw it to the attention of the architect.

That position was taken a stage further by a Court of Appeal case.[63] Although this case involved sub-contract work, the principles set out by the court are equally applicable to main contracts. JMH designed the temporary support work to a roof. Unfortunately, its design was overruled by the employer's engineer, who proposed a different design. There was no question in this instance over whether JMH failed to warn the engineer. They did warn the engineer of the danger of his design quite clearly, but he took no notice and the engineer's design for temporary work went ahead. Needless to say, the roof collapsed. Surprisingly, the court held, not just that JMH had a duty to warn, which the court seemed to accept had been done, but that they had failed to warn with sufficient force. One cannot help but think that the only degree of warning that the court would have accepted as sufficient would have been if JMH had given the warning and, at the same time, threatened to stop work if the warning went unheeded. This appeared to be the court's position also.

The contractor's duty to warn probably arises only if the design is seriously defective. In the case just mentioned, it seems to have been a potential danger to life. A contractor who did not warn an architect who had made a small dimensional error or a small mistake in detailing would be unlikely to have any liability.

The important point to be drawn from these cases is the reliance by the employer on the contractor. If it can be shown that the employer does rely, even partly, on the contractor, it seems that there will be a duty to warn of serious defects. On the other hand, cases where the duty arises to warn the architect will be rare, because the architect seldom, if ever, relies or is entitled to rely on the contractor. In the

61 *University Court of the University of Glasgow v William Whitfield & John Laing (Construction) Ltd* (1988) 42 BLR 66.

62 *Edward Lindenberg v Joe Canning & Jerome Contracting Ltd* (1992) 29 Con LR 71.

63 *Plant Construction plc v Clive Adams Associates and JMH Construction Services Ltd* [2000] BLR 137.

context of JCT traditional contracts, the duty is likely to be limited, because the employer will usually be relying on the architect and not the contractor. Contractors can take heart that they are not generally responsible for checking the architect's drawings. Having said that, a contractor proceeding with construction in the certain knowledge that there were errors on the drawings would find little favour with an adjudicator in any subsequent dispute.

40 Under DB, must the employer's agent approve the contractor's drawings?

Clause 2.8 provides that the contractor must without charge give the employer two copies of its design documents as and when from time to time necessary and in accordance with schedule 1 of the contract or as otherwise stated in the contract documents. The contractor is not to commence any work until it has complied with the procedure.

Schedule 1 sets out the procedure, but with reference to the Employer's Requirements. Paragraph 1 requires submission in the format stated in the Employer's Requirements. Therefore, if the Employer's Requirements do not state the format, it seems the contractor may submit the information in any format it desires. It might even be argued that, in the absence of a stated format, the contractor effectively need not submit at all. That is a very strict view, but one which a contractor is entitled to take. Therefore, it is essential that the format is set out.

The submission must be made in sufficient time to allow any comments made by the employer to be incorporated before use of the relevant document. That must be read in the context of paragraph 2, which gives the employer 14 days from receipt of the submission or, if the contract documents give a later date or a period, from the date or the expiry of the period, to return one copy of the document to the contractor.

The contract adopts the well-known system of lettering the returned documents either 'A', 'B' or 'C', depending on whether or not they are in accordance with the contract. 'A' means that the contractor must carry out the Works in accordance with that document. 'B' or 'C' means that the document is not in accordance with the contract and it must be accompanied by a written statement stating why the employer considers that to be the case. Documents marked 'B' may be used by the contractor if the employer's comments are incorporated and the employer is provided with an amended copy.

Documents marked 'C' cannot be used for construction, but the contractor may resubmit after amendment.

If the contractor thinks that the employer is wrong and that the document is in accordance with the contract, there is the option under paragraph 7 of notifying the employer within 7 days of receipt of the comment that compliance with the comment will result in a change (i.e. a variation). The contractor must give a reason, of course. The employer has a further 7 days to either confirm or withdraw the comment. If the employer simply confirms the comment, the contractor must then amend and resubmit the document. Paragraph 8 then sets out some provisos:

- Whether the employer confirms or withdraws comments does not mean that the employer accepts that the documents or amended documents are in accordance with the contract or that compliance with the comments will result in a change.
- If the contractor does not take the option of notifying the employer that compliance with the comment will result in a change, the comment is not to be treated as giving rise to a change.
- The contractor's duty to ensure that the design documents are in accordance with the contract is not reduced by the contractor's compliance with the submission procedure or with the employer's comments.

In brief, the position is that it is the contractor's obligation to comply with the contract. No submission of documents or comments by the employer will remove that obligation. If the employer makes comments that amount to a change, the contractor must promptly notify the employer of its view on the matter. Failure to notify the employer within the 7 days allotted will preclude the contractor from recovering any payment for such alleged change. However, notification, in itself, will not guarantee payment; it will be a matter of fact whether or not there has been a change.

It should be noted that the contract stays well clear of any suggestion that the employer approves any documents. But use of the word 'approval' appears not to make any difference to the principle in any event. In *Hampshire County Council v Stanley Hugh Leach Ltd*,[64] the court said:

64 (1991) 8-CLD-07-12.

The fact that Leach's alternative proposals were approved by the architects is irrelevant. No employer is going to be advised to enter into a contract giving the contractor an entirely free hand. The JCT Design and Build Contracts require the contractor's design be approved and this of course does not relieve the contractor of obligations in respect of his design.

41 What happens if the contractor cannot obtain materials?

The authority on this topic is scarce to the point of non-existence.

Under SBC clause 2.3.1, materials and goods have to be provided only 'so far as procurable'. It will be noticed, however, that the contractor's obligation under clause 2.1 is to provide what is specified in the contract documents. The whole of the contract must be read together, of course, and the introduction of the word 'procurable' gives the contractor a useful protection if materials or goods are truly unobtainable. Clearly, it does not protect a contractor who discovers that it has miscalculated its tender and that it is more difficult or more expensive than expected to provide what is specified. 'Procurable' is not qualified and, on a strict reading of the clause, it can even be argued that a contractor is protected even if the materials or goods were not procurable before the contract was entered into. It might be thought that a sensible and businesslike approach would restrict the meaning of 'procurable' to those items that had become unobtainable after the contract was executed. Whether that is the correct way to interpret it is not certain.

It is difficult to forecast what conclusion might be reached by the courts, still less an adjudicator, but a strict reading of the clause results in the conclusion (deeply unattractive so far as the employer is concerned) that if the items are not procurable for any reason, the contractor's obligation to provide them is at an end. It then becomes necessary for the architect to issue an architect's instruction requiring as a variation the provision of a substitute material. The variation is to be valued in the usual way.

The effect is to remove from the contractor any obligation to check that specified goods and materials are procurable before tendering. In order to change this situation, it is probably necessary to amend the contract clauses to specify a date after which the contractor is not responsible if materials or goods are no longer procurable.

It is arguable that if the contract does not refer to materials being procurable, the contractor's inability, through no fault of its own, to obtain specified materials may render the contract frustrated.

42 What if the contractor argues that the standard of work should take into account the tender price and a lower standard should be expected if the price is low?

A contractor may occasionally argue in this way if it is consistently failing to achieve a good standard, irrespective of the actual price. SBC clause 2.3.3 provides that where and to the extent that quality of materials or standards of workmanship is a matter for the architect's opinion, they are to be to the architect's reasonable satisfaction. It has been held that all work and materials are inherently matters for the architect's satisfaction.[65] Therefore, the question can be far reaching in its consequences. The contractor's argument is based on an old case concerning a dispute between architect and employer.[66] The architect claimed outstanding fees and the employer counterclaimed, alleging negligence on the part of the architect in failing to exercise due skill and care and ensuring that the house was properly constructed. The court held that there was some poor material and poor workmanship but, taking into account that the house was being built down to a price, a certain tolerance must be expected and the architect had not failed as alleged. This view was confirmed by the Court of Appeal although Lord Denning dissented.

It is quite difficult to understand the reasoning behind the judgment. The best that can be said is that it concerned disputes between architect and employer and not between employer and contractor. Therefore, it has little, if any, application to dispute with the contractor. The law is clear. If the contractor has undertaken to carry out work to a certain standard, that is the standard required and if a lower standard is provided, the contractor will be in breach of contract. The level of the tender price, which becomes the contract sum, is not relevant other than that is the price to be paid for the work as specified. In the case noted above, the specification called for the workmanship and materials to be the best of their kind and to the

65 *Crown Estates Commissioners v John Mowlem & Co Ltd* (1994) 70 BLR 1.
66 *Cotton v Wallis* [1955] 1 WLR 1168.

satisfaction of the architect. That is a very high standard. The best does not mean second best. Importantly, there was no reference at all to the pricing level. For example, the contract did not say that the standards should be reduced in view of the low contract sum. If I ask the price of a two year old car and I am told that it is £5,000 and I accept, the seller is not entitled to give me a four year old car on the ground that £5,000 is very cheap for a two year old car and, therefore, I should be satisfied with something older. Exactly the same principle applies to the standard of work in a building contract. The employer is entitled to receive what has been agreed in the contract. In any event, it is not at all clear how, except in the most extreme of cases, anyone can form the view that the contractor's price is so low as to indicate a lower standard than specified. In the case of every tendering process, there will be a lowest tenderer who could presumably argue that its price, being lower than the rest, represents a lower standard of work. If that was to be the case, chaos would ensue. Therefore, a contractor trying to excuse poor work with that kind of argument has little chance of success.

43 What powers does a project manager have in relation to a project?

Over the last few years the concept of project management has steadily gained ground, together with a good many misconceptions. A project manager is unlikely to be the same person as the contract administrator and it is the contract administrator who has the main powers under the building contract. The RIBA approved a definition of a project manager as follows:

> The Project Manager is a construction professional who can be given *executive authority and responsibility* to assist the client to identify the project objectives and subsequently supply the technical expertise to assess, procure, monitor and control the external resources required to achieve those objectives, defined in terms of time, cost, quality and function.[67]

It may be argued that such a definition does nothing to set out what ought to be the function of a project manager in regard to a building

67 Project Management – A role for the Architect (June 1995) RIBA Practice Committee paragraph 32.

project. Project management is often considered as though it is a self-contained system, and the words 'project manager' instantly conjure up a recognisable and easily identifiable discipline. Even the courts have agreed that this is not the case and that the duties of a project manager may vary dependent on the base discipline of the person carrying out the role.[68] The concept of project management is not particularly linked to construction; a project manager is rather a creature of the manufacturing industries. It is certain that all project managers have skills in common, but a project manager on a building contract cannot approach the task with the same freedom as if he or she were project managing a new product through a factory. There are roughly two kinds of project managers:

- project managers who represent the employer and act as its technical arm;
- project managers who not only represent the employer but also carry out the contract administration role in regard to building contracts.

The first type of project manager acts as the employer's representative and generally acts as agent for the employer with the power to do everything the employer could do in relation to the project. This is probably the usual position occupied by the person termed project manager. He or she will appoint consultants and carry out the briefing exercise having first been briefed by the employer, make the final decision about the selection of a building contractor and answer any queries from the professional team. The theory is that the employer has the benefit of a skilled professional looking after his or her interests and being paid to watch the other professionals. This kind of project manager has no powers under the building contract although he or she may try to enter site, chair site meetings and give instructions directly to the contractor. Some project managers even insist on countersigning all certificates. That kind of activity on the part of the project manager, though regrettably common, is unlawful and it may lead to disputes. It always leads to confusion. A contractor taking instructions from such a project manager is most unwise.

Most building contracts do not even acknowledge the project manager's existence. There is provision under SBC for an employer's

68 *Pride Valley Foods Ltd v Hall & Partners* (2000) 16 Const LJ 424.

representative, who might well be the project manager, to be appointed to carry out the employer's functions. Some other contracts, for example GC/Works/1 (1998) refer to the project manager, but they might as well have used the phrase 'contract administrator' and simply add to the current confusion over the role.

The second type of project manager performs all the functions of a contract administrator so far as the building contract is concerned. The project manager in this situation has a great deal of power, because acting as the employer's representative is added to the role. The project manager's function is commonly thought by employers to be the management of the project and this type of project manager is closest to that situation. It is fairly unusual to find a project manager in this role. That is possibly just as well, because it is tantamount to having the employer as contract administrator and there is no properly independent professional for the issue of certificates. That very much devalues any certificates and makes them no more than the employer's view, which carries no more weight than the contractor's view.[69]

It is possible to find a project manager working solely for a contractor. In such cases, the project manager has no more, although different, power than a project manager acting solely for the employer. Indeed, it is useful to compare them.

44 Can certificates and formal AIs be issued if the contract is not signed?

This is a question that often troubles architects when work has begun under a so-called letter of intent, a month has gone by and the contractor is yelling for a certificate. Strangely, the architect has probably issued several Architect's Instructions by this time. It is the request for money which, as usual, concentrates the mind.

Obviously, contracts should always be signed by both parties before any work begins on site. The use of letters of intent is not to be encouraged, because they lead to a false sense of security. If the parties had contracted on what might be termed a simple contract where the contractor agreed to carry out work for a price and start and completion dates were agreed, the architect would have no power to issue either certificates or instructions. Indeed, an architect in these

69 *JF Finnegan Ltd (No 1) v Ford Sellar Morris* (1991) 25 Con LR 89.

circumstances would have no power at all because there would be no mention of the architect in the simple contract. Certain clauses would be implied by statute or under the general law, but the presence of an architect, or a quantity surveyor for that matter, would not be one of them.

If work is being carried out under a true letter of intent, a very limited contract would be formed of the 'if' variety: 'If you do some work, I will pay you a reasonable amount of money.' But few, if any, other terms would be implied and certainly the architect would have no rights or obligations under it.

However, the situation may be that the contractor has been invited to tender on the basis of drawings and specification or bills of quantities and these documents may include the clearest details of the contract to be executed, including how all the contract particulars will be completed. If the contractor submits an unqualified tender on that basis and if the employer proceeds to accept the tender without any equivocation, a binding contract will be formed incorporating all the details of the drawings and other documents in the invitation to tender and, most importantly, incorporating the terms of the contract specified in the documents. The architect will then be able to act exactly as if the parties had executed the formal contract documents.

Of course, if the acceptance or the invitation to tender makes reference to acceptance being subject to the execution of formal documents, no contract is in place until that is done. On the other hand, tenders may be submitted with qualifications or letters of acceptance may include conditions that make them counter-offers, but the qualified tender may be unequivocally accepted or the counter-offer may be accepted by the contractor and a binding contract come into existence in that way. The possible permutations are probably endless and great care is required to properly categorise the relationship.

45 Is the architect obliged to check the contractor's setting out if requested?

It is quite common for the contractor to request the architect to check the setting out and it is usually prudent for the architect to do so. SBC clause 2.10 provides that the architect must determine any levels required for the execution of the Works and must supply accurately dimensioned drawings so that the contractor can set out the Works. The contractor is responsible for correcting errors in its own setting out at no cost to the employer, but that is little consolation to an employer who is facing legal proceedings for trespass from a neighbour for a building which is almost finished and encroaches on neighbouring land: particularly if the contractor chooses that moment to go into liquidation.

There is nothing in the contract which places a duty on the architect to check setting out and the architect owes no duty to the contractor to do so. However, the architect certainly owes a duty to the client to inspect the Works. Although that duty does not extend to checking every last detail of the Works, the architect must inspect all the important aspects of the building work. What could be more important than the setting out of the building on the site? It is dangerous for the architect to confirm to the contractor that the setting out is accurate, because it tends to relieve the contractor of responsibility for errors which it would otherwise have under the contract. The architect should not wait for the contractor to request checking of setting out. A visit to the site and to check the setting out should be a priority. Contractors often write to the architect, confirming that its setting out was found to be accurate. A wise architect will immediately respond, making quite clear that no approval is given to the setting out and reminding the contractor of its responsibilities under the contract.

46 If the architect finds that there is no person-in-charge on site, can the project be halted until the person-in-charge is on site?

SBC and DB clause 3.2 require the contractor to have a person-in-charge on site 'at all times'. IC, ICD, MW and MWD clause 3.2 require the person-in-charge to be on the site 'at all reasonable times'. The words 'at all times' is not to be taken literally and it is suggested that under SBC and DB, the contractor's obligation amounts to ensuring that the person-in-charge is on site for the whole of the time that any activity is being carried out, even if that is simply off-loading materials. The JCT draftsman's decision to use different wording for the Intermediate and Minor Works contracts clearly suggests a somewhat lesser duty under those contracts. What is reasonable will depend on all the circumstances. Generally it is likely that the person-in-charge must be on site whenever the Works are in progress although there may be some instances, where the Works are small or carried out in small parcels over a long period, which do not warrant that the person-in-charge is constantly on the site. It may be sufficient to visit the site on a regular basis provided that there is a responsible operative available. On small projects, it is common to have a working foreman and for the identity of the foreman to change as the work progresses to reflect the progression in trades.

Therefore, the first thing is to establish whether, under the particular contract terms, the person-in-charge should be on site. The second thing is to check how long the site has been left without supervision. It may be that the person-in-charge has simply left site on an urgent matter for a few minutes and will shortly return. It will be serious, for example, if the architect arrives on site and finds that the person-in-charge has been away all morning. If the person-in-charge is away from the site during the time that work is in progress, it is certainly a breach of contract under SBC and DB and possibly so, depending on circumstances, under IC, ICD, MW and MWD. The contract does not include any direct sanction for such a breach and if the project simply continued despite the absence of the person-in-charge, the employer would be left to claim whatever damages he or she had suffered as a result of the breach. That would generally amount to the cost of correcting any defects in the work. If despite the absence of the person-in-charge, there were no defects, it is difficult to see what damage the employer could be said to have suffered. The employer would always have the concern that, during such absence, some essential work, now covered up, had been badly done or even omitted, but suspicion without evidence would not give rise to damages and it is a common experience that defects occur even when a person-in-charge is constantly on site.

Under SBC clause 3.15, DB clause 3.10 and IC and ICD clause 3.12 there is power for the architect (the employer under DB) to issue instructions postponing any work and the power is probably implied in MW and MWD clause 3.4. Usually, when the power is exercised, it almost automatically entitles the contractor to an extension of time and appropriate loss and/or expense. The position under MW and MWD is less clear so far as loss and/or expense is concerned. However, it seems likely that an architect who postpones the work as a direct result of the absence of the person-in-charge would be justified in withholding both an extension of time and loss and/or expense which would otherwise be due. It seems to be a sensible exercise of the architect's powers to postpone the work pending the return of the person-in-charge to site after sending a warning letter to the contractor, reminding it of its contractual responsibilities. In practical terms, the project cannot proceed if the person responsible for organising and supervising the work is not present on site.

47 If the employer sacks the architect under MW and appoints an unqualified surveyor as contract administrator, is the contract still valid?

The identity of the architect is stated in article 3 of the JCT Minor Works Building Contract 2005 (MW). It is similar to its predecessor (MW 98) in that it states that if the architect ceases to act, the replacement architect will be the person nominated by the employer within 14 days. There is a proviso that the replacement architect must not disregard or overrule any certificate, opinion, decision, approval or instruction given by the former architect except to the extent that the former architect would have been able to do so.

In article 3, reference is to the 'Architect/Contract Administrator' and this way of describing the person named in article 3 is adopted throughout the contract. Footnote [7] of MW explains that if the person named in the contract is entitled to use the name 'Architect' in accordance with the Architects Act 1997, 'Contract Administrator' should be deleted and is then deemed deleted throughout the rest of the contract. The purpose of providing the alternative name is to protect an unregistered person from being prosecuted under the Act.

It follows, therefore, that, if the person originally named in the contract is an architect, the words 'Contract Administrator' must be deleted or if not deleted will be deemed deleted there and throughout the contract. Clearly, an unregistered person, even if a qualified surveyor, cannot be appointed as architect. Such an appointment would not invalidate the contract, but it would be unlawful and of no effect. Indeed, the person concerned would be liable to prosecution for infringing the Act and may have to pay a substantial fine.

There is another, separate, consideration. When the contractor tendered, it would have been on the basis that the contract would be administered by an architect of known ability. It is a sound argument that, in the case of any replacement, the replacement person must be of the same ability. This would prevent the all-too-prevalent practice whereby an employer sacks the architect and self-appoints.

If the position was reversed and the unqualified surveyor was sacked and replaced with an architect, there should be no difficulty, because the surveyor would have been described throughout the contract as the 'Contract Administrator' and the architect certainly fits into that category.

Under the JCT Standard Building Contract 2005 (SBC), article 3 states the name of the architect and clause 3.5 provides that renomination must take place within 21 days of the architect ceasing to hold the post. The contractor is given express power to object within 7 days for a reason accepted by the employer or considered to be sufficient by an adjudicator, arbitrator or judge unless the employer is a local authority and the former architect was an official of it. The JCT Intermediate Building Contract 2005 (IC) names the architect in article 3 and provides for replacement in clause 3.4.1 in a somewhat shortened version of the SBC clause but to similar effect, save that the employer has only 14 days to nominate.

48 SBC: There is a clause in the bills of quantities preliminaries which states that no certificates will be issued until the contractor has supplied a performance bond. Work has been going on site for 6 weeks and there is no performance bond, but the contractor says that the architect must certify. Is that correct?

This is a surprisingly common provision. The contractor is correct. In *Gilbert Ash (Northern) v Modern Engineering (Bristol)*,[70] this kind of action was held to amount to a penalty, that was therefore unenforceable, because large amounts of money could be withheld for a trivial breach. It can readily be seen that if a performance bond is required in the amount of 10 per cent of the contract sum and if 10 per cent of the contract sum was, say, £85,000, the contractor might well have earned this amount in 2 or 3 months. Therefore, to withhold payment beyond that point would be to penalise the contractor unduly. This is something the courts have never condoned. In considering liquidated damages and penalties, the courts have made clear that a greater sum can never be proper recompense for the loss of a lesser sum.[71]

The Court of Appeal took a down-to-earth view in relation to a contract's commercial aspect in a more recent case involving the

70 [1973] 3 All ER 195.
71 *Kemble v Farren* (1829) 6 Bing 141.

Millennium Dome.[72] The contract between Koch and Millennium (for the supply and fixing of the roof) contained a clause which said: 'As a condition precedent to any liability or obligation of the client under this Trade Contract, the Trade Contractor shall provide at its own costs a guarantee in the form outlined in Schedule . . .' The contract documents were completed by Koch, but no guarantee and performance bond was completed. Koch confirmed that a guarantee and performance bond would be completed and sent. Koch then heard that the contract might be given to another company and suspended the execution of the guarantee and bond until the position was clear. Subsequently Koch's employment under the trade contract was terminated. Millennium then argued that it was not obliged to make any payment to Koch, because the condition precedent was not satisfied.

When the matter came before the Court of Appeal, the court thought that Millennium's contention was misconceived. The purpose of the guarantee and bond was to ensure that Millennium was protected when the works commenced. The court stated:

> It was suggested on behalf of the Millennium Company, that the purpose is achieved by relieving the client from the obligation to make any payments until the guarantee and the performance bond have actually been provided. But, as it seems to me, the client and the trade contractor cannot have intended that the effect of their agreement should be that the trade contractor should be entitled to carry on works without being paid for some indefinite period until it chose to provide the guarantee and performance bond. Such an arrangement could properly be described, in my view, as commercial nonsense.

In addition, by choosing to terminate Koch's employment under the contract, Millennium made it impossible for the condition precedent to be fulfilled.

This case makes clear that the courts will take a dim view of very onerous conditions in business contracts if they do not make commercial sense.

72 *Koch Hightex GmbH v New Millennium Experience Company Ltd* (1999) CILL 1595.

49 Funding has been stopped and three certificates are unpaid. The contractor has suspended obligations. Subsequently, vandalism occurred on site – whose problem is that?

Section 112 of the Housing Grants, Construction and Regeneration Act 1996 provides that if a sum due under a construction contract is not paid in full by the final date for payment and no withholding notice has been served, the person to whom the sum is due has the right to suspend performance of all obligations under the contract.

A contractor engaged on a construction contract that does not have a residential occupier is entitled to enforce that right by giving at least 7 days' written notice to the employer. The right is included in SBC by clause 4.14, in IC and ICD by clause 4.11 and in MW and MWD by clause 4.7. The principal difference between the JCT contracts and the Act is that each of the JCT clauses requires the contractor to send a copy of the suspension notice to the architect. If that is not done, the suspension is still valid, but it is carried out under the Act rather than under the contract terms. Where a copy is sent to the architect, the suspension is carried out under the terms of the contract that provide, in the case of SBC, IC and ICD, that the contractor will be entitled to an appropriate extension of time and any loss and/or expense caused by the suspension. The Act provides only for extension of the contract period.

In this question, it seems that the contractor has exercised a remarkable – one could say foolish – degree of forbearance so far as the three unpaid certificates are concerned. It is presumed that the requisite 7 days' written notice has been given; if not, then the situation would be less clear. Failure to pay one certificate is a breach of contract on the part of the employer, which entitles the contractor to take steps to terminate the contract under the appropriate clauses (SBC, IC and ICD: clause 8.9, MW and MWD: clause 6.8). Failure to pay three certificates probably entitles the contractor to accept the employer's conduct as repudiatory, bring the obligations of both parties to a permanent end and claim damages.[73] But if there has been no acceptance of the repudiation and the contractor has merely suspended its obligations without giving the necessary 7 days' notice, it amounts to a breach of contract on the part of the contractor. It is

73 *DR Bradley (Cable Jointing) Ltd v Jefco Mechancial Services* (1989) 6-CLD-07-19.

probably not a breach that entitles the employer to accept it as repudiation for two reasons:

- The contractor is not saying that it will never continue its obligations, but simply that it will suspend them until it is paid. An expressed intention to suspend precludes the implication of an intention to bring the contract to an end.[74]
- The contractor intends to comply with the contract, albeit it has gone about this improperly, therefore there is no intention to repudiate the contract.[75]

Nevertheless, it is a breach of contract and the employer would be entitled to such damages as flow from the breach. The cost of rectifying the acts of vandalism might well fall into that category.

The position is totally different if the contractor has given the requisite notice. When the notice expires, the contractor is entitled to leave the site. The contractor owes a general duty to leave the site in a safe condition, but the duty to insure comes to an end together with all the contractor's other contractual duties. If vandalism occurs, the failure of the employer to make payment is not the cause, but it is a circumstance without which the vandalism probably would not have occurred. The reason the site was unprotected was because the contractor had exercised the right to suspend. The problem would be the employer's problem, along with all the other problems associated with total cessation of work on site.

50 Is it impossible to say that a contractor is failing to proceed regularly and diligently?

A few years ago, it used to be said that it was not possible to say that the contractor was failing to proceed regularly and diligently while ever there was anyone at all on the site, even though the pace of the work was snail-like. Everyone knew that it was just plain silly to take that attitude but there was great reluctance on the part of architects to take any action on the basis of failure to proceed regularly and diligently, because of the fear of getting it wrong. SBC, IC, ICD clause 2.4 and DB clause 2.3 use this phrase to describe the contractor's

74 *F Treliving & Co Ltd v Simplex Time Recorder Co UK Ltd* (1981) unreported.
75 *Woodar Investment Development v Wimpey Construction UK* [1980] 1 All ER 571.

duty to progress. Because breach of this duty is a ground for termination under SBC, DB, IC, ICD, MW and MWD it is important to be able to identify it. Even the courts were doubtful about its precise meaning one judge concluded his consideration of the topic with the words: 'All I can say is that I remain somewhat uncertain as to the concept enshrined in these words,' which was not very helpful.[76] The position changed abruptly when it was held that architects were negligent for not issuing a default notice to a contractor specifying that it was not proceeding regularly and diligently. The Court of Appeal turned down the architects' appeal and laid down some useful guidelines:

> Although the contractor must proceed both regularly and diligently with the Works, and although each word imports into that obligation certain discrete concepts which would not otherwise inform it, there is a measure of overlap between them and it is thus unhelpful to seek to define two quite separate and distinct obligations.
>
> What particularly is supplied by the word 'regularly' is not least a requirement to attend for work on a regular daily basis with sufficient in the way of men, materials and plant to have the physical capacity to progress the work substantially in accordance with the contractual obligations.
>
> What in particular the word 'diligently' contributes to the concept is the need to apply that physical capacity industriously and efficiently towards the same end.
>
> Taken together the obligation upon the contractor is essentially to proceed continuously, industriously and efficiently with appropriate physical resources so as to progress the works steadily towards completion substantially in accordance with the contractual requirements as to time, sequence and quality of work.
>
> Beyond that I think it impossible to give useful guidance. These are after all plain English words and in reality the failure of which [the clause] speaks is, like the elephant, far easier to recognise than to describe.[77]

76 *London Borough of Hounslow v Twickenham Garden Developments Ltd* (1970) 7 BLR 81.

77 *West Faulkner v London Borough of Newham* (1995) 11 Cons. LJ 157 (CA).

The final sentence quoted is probably the most telling. The number of operatives, the amount of plant and equipment and the organisation of the site are important factors in considering whether or not the contractor is progressing regularly and diligently. The contractor's progress compared with the programmed progress is significant, but not conclusive. Whether the contractor is going ahead regularly and diligently is to be judged by the standards expected of the average competent and experienced contractor. The contract administrator is in the best position to judge whether work is proceeding regularly and diligently and there is a clear duty to advise the employer and warn the contractor accordingly.

Chapter 8

Architects

51 Planning permission was obtained for a small building. The building owner wants to press ahead with a larger building without further reference to Planning. The architect knows that the Planning Department would refuse the large building out of hand. Should the architect continue to do the drawings and administer the contract on site?

It is the architect's duty to advise the client on all aspects of the building process about which the architect professes expertise. Obviously, town planning is one area where the architect should have expertise – not the expertise of a town planner or of an expert planning consultant or of a lawyer skilled in this area of the law, but certainly the ordinary aspects of town planning that one would normally expect the architect to know as part of the general architectural skills.[78]

In this case, planning permission had been obtained for a particular building. The client had clearly had a change of mind and wanted a bigger building on the same site. There is nothing wrong with that, but architects have a duty to advise the client about seeking planning permission again in such circumstances. If intending to press ahead without obtaining planning permission, the client is going to do something unlawful and the architect has a duty to make that crystal clear. In addition, if the architect knows for certain that making a planning submission would be pointless, there is an added duty to convey this to the client in the strongest terms.

The architect is being asked to finish the drawings for the larger building that would not gain planning approval even if application

78 *BL Holdings v Robert J Wood and Partners* (1979) 12 BLR 1.

were made. The architect knows that the purpose for which the drawings are intended is the unlawful construction of a building. Architects placed in this position should flatly refuse to have anything further to do with the project. This would not amount to repudiation at law, much less be in breach of any part of the professional Code of Conduct. Quite the contrary: the larger building is not something for which the architect was engaged. In other words, it is not part of the terms of engagement (for the smaller building for which planning permission has been obtained) and, therefore, the architect cannot be in breach by refusing to do the work.

For an architect to collaborate with the client in enabling the construction of a building for which it is known that planning permission would be refused is an unlawful act and one that is contrary to the Code of Conduct. If the architect did not know that planning permission was not to be sought or that it could not be obtained, there would be no wrongful behaviour until the architect knew, or should have known, the true situation. On becoming aware of the true situation, if advice to the client fell on deaf ears, the architect would have no option but to stop work.

52 Are there any circumstances in which a contractor can successfully claim against the architect?

This is a question that crops up fairly frequently. Architects are prone to ask it just before making an important contractual decision; contractors ask it when they are particularly annoyed with an architect's conduct. In general terms, it is usually easier for the contractor to claim against the employer than the architect. That is because the contractor and the employer are related by the building contract while the contractor has no contractual relationship with the architect. Therefore, a contractor finds it relatively easy to claim under the terms of the contract or even against the employer for a breach of the contract. Claims by the contractor against the architect have to be in tort. Since the case of *Murphy v Brentwood District Council*,[79] negligence claims have become very difficult to sustain. A contractor making a claim against an architect would almost certainly do so under the reliance principle.[80] It usually applies to professionals,

79 (1990) 50 BLR 1.
80 *Hedley Byrne & Co v Heller & Partners Ltd* [1963] 2 All ER 575.

although courts have extended the scope in some instances. The principle, in brief, is to the effect that if a professional gives advice to another person or class of people knowing that the person or persons will rely on it, and if the person or persons does rely on the advice and, as a result, suffers a loss, the loss will be recoverable from the professional. This is irrespective of any fee paid or not paid to the professional and even if there is no contractual relationship.

The contractor sued both employer and architect in *Michael Salliss & Co Ltd v ECA Calil,*[81] claiming that the architects owed a duty of care to the contractor. Although the contractor was unsuccessful in arguing that the architect owed the contractor a duty to provide accurate and workable drawings, it was successful in its claim that it relied on the architect to grant an adequate extension of time and properly to certify the value of work done. The court appeared to think that it was self-evident. It remarked that, if the architect unfairly promoted the employer's interest by inadequate certification or merely failed properly to exercise reasonable care and skill in the certification, it was reasonable that the contractor should not only have rights against the owner but also against the architect to recover damages.

Three years later, *Pacific Associates v Baxter*[82] seemed to overturn this position, but although it was said that a question mark hung over the *Sallis* case, it stopped short of saying that it was wrongly decided. Pacific Associates was effectively the contractor under a FIDIC contract for work in Dubai. The contractor claimed that it had encountered unexpectedly hard materials and that it was therefore entitled to a substantial extra payment. The engineers would not certify the amount claimed and the contractor sued them. The claim alleged that the engineers acted negligently in breach of their duty to act fairly and impartially in administering the contract. In a judgment upheld by the Court of Appeal the court struck out the claim, on the basis that the contractor had no cause of action or, in other words, that it could not make a claim in that particular way. In making that decision, the court was mindful that there was provision for arbitration between employer and contractor and, therefore, the contractor could have sought arbitration on the dispute.

81 (1987) 13 Con LR 68.
82 (1988) 44 BLR 33.

The court also referred to a special exclusion of liability clause in the contract by which the employer was not to hold the engineers personally liable for acts or obligations under the contract, or answerable for any default or omission on the part of the employer. The fact that the engineers were not a party to the FIDIC contract – just as architects are not parties to JCT contracts – seems to have been ignored by the court.

The question mark appears to properly hang over the *Pacific Associates* case rather than the *Sallis* case. Whether a duty of care exists does not depend on the existence of an exclusion of liability clause, except to the extent that the existence of such a clause suggests acceptance by the engineer that there is a duty of care which, without such a clause, would give rise to the liability. The clause in question might well be deemed unreasonable under the provisions of the Unfair Contract Terms Act 1977.[83] Why the inclusion of an arbitration clause should exclude engineers from liability to the contractor is not immediately (or even subsequently) obvious. The fact that the parties chose to settle their disputes by arbitration cannot excuse the engineers from their duty to both parties.

In *Lubenham Fidelities & Investment Co v South Pembrokeshire District Council and wigly fox Partnership,*[84] the Court of Appeal (by which the court in the *Pacific Associates* case should have been bound) expressly confirmed that the architect owed a duty to the contractor in certifying. The architects in that case were not held liable, but that was because the chain of causation was broken and the contractor's damage was caused by its own breach in wrongfully withdrawing from site. But the court reached its conclusion with reluctance, because the architects' negligence was the source from which the sequence of events began to flow. The court expressly stated that, because the architects were appointed under the contract, they owed a duty to the contractor as well as to the employer to exercise reasonable care in issuing certificates and in administering the contracts correctly and that, by issuing defective certificates and in advising the employer as they did, the architects acted in breach of their duty to the contractor.

The court was simply following the precedent of earlier courts. In *Campbell v Edwards,*[85] the Court of Appeal said that contractors

83 *Smith v Eric S Bush* [1989] 2 All ER 514.
84 (1986) 6 Con LR 85.
85 [1976] 1 All ER 785.

had a cause of action in negligence against certifiers and valuers. Until *Pacific Associates* it had not been doubted that architects owed a duty to contractors in certifying. In *Arenson v Arenson*,[86] in reference to the possibility of the architect negligently under-certifying, it was said that, in a trade where cash flow is perceived as important, it might have caused the contractor serious damage for which the architect could successfully have been sued. In *F G Minter Ltd v Welsh Health Technical Services Organisation*,[87] the court remarked that an unreasonable delay in ascertainment would completely break the chain of causation, which might give rise to a claim against the architect.

More recent cases[88] provide firm support to the idea that the reliance principle established in *Hedley Byrne* can be extended to actions as well as advice given by the architect.

53 If the contract requires an architect to 'have due regard' to a code of practice, does that mean the architect must comply with it?

A requirement to have due regard to something is a common phrase in much legislation and in some contracts. It must be construed in the context of the surrounding words. For example in SBC clause 3.18.4 if any work, materials or goods are not in accordance with the contract, the architect is required to have due regard to the code of practice attached to the contract before issuing instructions for opening up the work or testing. A requirement to have due regard to something is not the same as a requirement to comply with something. In SBC clause 3.10, the contractor must comply forthwith with architect's instructions. In other words, the contractor must carry out the instructions. It would not be sufficient if the contractor merely had to have due regard to the instructions. A requirement to have due regard to something means that a person must carefully consider that thing and give appropriate weight to it in coming to a decision or in taking some action.

86 [1975] 3 All ER 901.

87 (1979) 11 BLR 1; partly reversed by the Court of Appeal, but not on this point (1980) 13 BLR 7 (CA).

88 *Henderson v Merritt Syndicates* [1995] 2 AC 145, (1994) 69 BLR 26; *White v Jones* [1995] 1 All ER 691; *Conway v Crow Kelsey & Partner* (1994) 39 Con LR 1; *J Jarvis & Sons Ltd v Castle Wharf Developments & Others* [2001] 1 Lloyd's Rep 308.

The code of practice referred to in SBC clause 3.18.4 is contained in schedule 4 to the contract. Its purpose is stated to be to assist in the fair and reasonable operation of the requirements of the clause. In order to have due regard to the code, an architect must read it and consider whether any of its contents were relevant to the decision to open up or test. However, there may be circumstances where the architect decides that no part of the code is relevant and, therefore, feels justified in giving no weight to it when making the decision.

54 Can an architect be liable for advising the use of the wrong form of contract?

There is little doubt that an architect can be liable for advising the use of the wrong form of contract. Several legal commentators have made much of it. However, it would have to be demonstrated that the employer suffered a loss as a direct result of using the form and that the loss was not such as would have been suffered in any event and quite irrespective of the type of contract. In other words, it would have to be shown by an employer claiming damages for the bad advice that it was the bad advice and nothing else which was the cause of the loss. In practice, it is likely that some architects do give their clients poor advice about a suitable form of contract for a particular project. Often, there will be no adverse consequence simply because circumstances do not arise which necessitate the use of the part of the contract which would disadvantage the employer. For example, an architect may wrongly advise an employer to use a particular contract which happens to have terms for termination of the contract which are particularly advantageous to the contractor. If there is no need to terminate, there are no consequences from the bad advice.

There is no doubt that the task of advising on the correct form of contract for a project is becoming more difficult. There are many different forms of contract available and a comparison of the key terms is not easy unless an architect has an encyclopaedic knowledge of all the available contracts and the individual clauses and their significance. Moreover, changes in legislation, judicial decisions and other factors mean that standard contracts are regularly updated. If in doubt, architects should seek advice on behalf of their clients or advise their clients to seek further advice themselves. Obviously, if an architect spends a career engaged in one specific kind of building design at a particular price range, it is likely that the same contract will be suitable for all the projects. The danger tends to arise when an architect uses the same forms of for all project simply because it is a familiar document. The author has come across many instances of architects who refuse to use other than the Minor Works Contract or sometimes the Intermediate Contract whatever may be the size or scope of the Works. That approach is asking for trouble.

55 The job went over time. The employer and the contractor did some kind of deal. Where does that leave the architect?

Employers and contractors are always doing some kind of deal. Often it concerns the final account; sometimes it is about extensions of time or liquidated damages; sometimes it concerns everything. This is particularly common among employers who are in business. They may say that time is more precious than money and, therefore, it suits them to reach a quick and relatively painless agreement even if they have to pay more.

What such employers forget is that they and the contractors concerned have entered into building contracts by which they have agreed that such things as extensions of time, loss and/or expense, liquidated damages and the amount due to the contractor will be decided by the architect. Obviously, it is always possible for the parties to a contract to decide on some change to that contract. If they decide that they prefer to sort out extensions of time rather than let the architect do it, that is a matter for them, but it also affects the architect.

Architects, by their terms of engagement, usually agree to administer the building contract in accordance with its terms. Contrary to the belief of some architects, an architect is only empowered to do that which the contract permits. Therefore, if an employer and contractor decide, for example, that the contract period will be extended by 6 weeks, they are entitled to do so, but that does not mean that the architect has to issue an extension of time for that period. In fact, the architect has no power under the contract to do anything in relation to the agreed extension. If the parties formalise it, the architect can probably regard it thereafter as the date for completion in the contract as varied by the parties. It is that varied date that the architect will have to consider in future if further delays are notified and further extensions may be due.

If the employer and the contractor agree that the contractor should be paid a sum of money as well as having the contract period revised, the architect cannot include the sum in any certificate, because it is not a sum properly due under the terms of the contract, but rather a sum agreed *ad hoc* by the parties. Therefore, it is for the employer to pay the sum directly to the contractor.

There are numerous little agreements and deals into which the parties can enter, leaving the contract terms relatively unscathed. However, certain deals may leave the architect in a difficult position. This can happen if the deals are done without notifying the architect or if the employer begins to interfere with the certification process, for example by insisting that the architect certify some extra agreed amount. The employer may enter into what are often euphemistically called 'acceleration agreements' with the contractor. These are often simply agreements that the contractor will work additional hours and weekends, not that the Works will progress faster.

Many architects finding themselves in this position will wonder where they stand, while desperately trying to administer the contract by adapting its terms to suit the problems that half-considered agreements can throw up. Architects should not do that. If the employer and contractor strike some deal that prevents the architect from properly administering the contract, it is probable that the employer has repudiated the architect's terms of engagement. In these circumstances, the architect should immediately seek expert advice on his or her position. If it is repudiation, it will entitle the architect to accept the repudiation, bringing the architect's obligations to an end and opening the door for a claim for damages against the employer.

56 What are the dangers in a construction professional giving a certificate of satisfaction to the building society?

The first thing to understand is that the certificate required by a building society is an entirely different thing to the certificate that an architect will routinely issue under a standard building contract.

The certificate, or more likely several certificates throughout the progress of the project, is required by the building society to give it insurance against the money it has been asked to lend. Where the builder is registered with the National House-Building Council, the architect will not usually be asked to give a certificate, because the guarantees of the NHBC are usually acceptable to the building society.

If, during or at the end of a project, the employer asks the architect to provide a certificate of satisfaction (sometimes referred to as a Professional Consultant's Certificate) and if, in the terms of engagement, the architect has not agreed to give such a certificate, there is no obligation to give one. Many architects, and their clients, will argue that if they have been engaged for a full service by their clients, they should be prepared to give the certificate at the end, because to do otherwise is tantamount to saying that they have no confidence in the work they have done. To take that point of view is to misunderstand the purpose and implications of giving such certificates. If an architect has been negligent in performing the services, the client can quite easily take legal action under the terms of the conditions of engagement. Whether such action takes the form of arbitration or legal proceedings through the courts depends on the terms of the engagement. The fact remains that the client has a perfectly adequate remedy for any default on the part of the architect and a separate remedy under the building contract for any defects in construction.

Why does the building society want a certificate before it will lend any money? The building society has an agreement with the architect's client, but it has no agreement with the architect. By completing the certificate, the architect is not only certifying that his or her own work has been performed properly, but also that the contractor's work has been carried out correctly in accordance with the building contract. At one time, architects used to be able to offer their own watered-down version of a certificate, in which their liability was very much restricted. More recently, building societies and banks have insisted on architects signing the building societies' and banks' own forms of certificates.

Sometimes architects do work for developers that are their own builders – in other words, a builder that has decided to build a few 'architect-designed homes' for the speculative market. Usually such a builder could obtain NHBC registration, but it is cheaper to ask the architect to provide a certificate of satisfaction.

The giving of such certificates is very dangerous for the architect, who is thereby exposed to an increase in the number of people who could successfully bring an action for damages, and it also significantly increases the liability as noted above. Architects giving such certificates know, or will be presumed to know, that the certificates will be used not only for the purposes of obtaining funding but also for selling the property to future purchasers, who might rely on the certificates in lieu of a building survey. This reliance goes to the very heart of the architect's liability. A building society or a future purchaser cannot bring an action against the architect in contract, because they have no contract with the architect. However, they can bring an action against the architect in tort on the basis of the certificates. Without the certificates, an action against the architect in tort for negligence would be very difficult to sustain.

The basis for the action is an old case called *Hedley Byrne & Co Ltd v Heller & Partners.*[89] Put very simply, if a professional person gives advice to a person or class of person knowing that the advice will be relied on and if the person receiving the advice does rely upon it and, as a result, suffers loss, the professional will be liable for such loss. See Question 52 for the ways in which this liability may be extended beyond professional advice in certain circumstances. Architects should refuse to give such certificates if not legally obliged to do so. If an architect does agree to give such a certificate, a substantial additional fee if indicated.

57 Is the client entitled to all the files belonging to construction professionals on completion of the project?

In theory, it depends on whether the professional is acting as principal by providing professional services to a client or whether the professional is acting as agent for a client. In practice these situations sometimes get mixed up. For example, architects often make planning applications or building regulation submissions as agents for their clients. Surveyors may attend meetings on behalf of clients or may be instructed by clients to negotiate boundaries and the like.

However, architects are not acting as agents when they give advice on the suitability of building sites or when they design buildings, administer contracts or act as expert witnesses. Quantity surveyors do not act as agents, but as principals, when they prepare bills of quantities or give cost advice.

If professionals are acting as principals, providing a service to their clients, such clients are entitled to what they have contracted to receive, but no more. For example, a quantity surveyor who has agreed to provide bills of quantities or specifications is obliged to provide such bills and specifications, but not any of the rough

89 [1963] 2 All ER 575.

calculations or workings that preceded them. An architect must provide designs and working drawings, but not the rough sketches and notes. If a client sends original documents, such as leases or conveyances, they obviously remain the client's property. If they are copies, it depends on circumstances whether they must be returned.

If professionals are acting as agents, their clients are entitled to all the documents created by the professionals or received by them from third parties in the course of discharging their duties.[90] For example, if an architect submits an application for planning permission, the client is entitled to the submission together with the approvals once received.

A professional is entitled to retain all the client's letters and copies of the professional's own letters to the client and all internal memoranda of the practice.[91] It is more likely that papers belong to the client when a professional is acting as agent, and more likely that they belong to the professional when that professional is providing a service.

Obviously, the parties to a contract can expressly agree to vary those general terms. Indeed, many terms of engagement that have been especially drafted for the benefit of the client have clauses that modify the position in favour of the client. Some are modified to such an extent that the client may call for any information required whether or not it is strictly part of the professional's own file.

58 When can job files be destroyed?

This question crops up again and again. Before the days of computer aided design, it used to be a particular worry of architects, who often had huge stocks of crumbling tracing paper, paper prints and slightly more durable linen or plastic film.

The basic answer to the question is that it is impossible to be sure that any particular file will not be required in the future. On that basis, no files should ever be destroyed. It is recognised that such a response is not really helpful for a professional, or anyone else for that matter, who is anxious to reduce the number of files on the shelf or in storage. The problem is best addressed as a process of elimination.

90 *Formica Ltd v ECGD* [1995] 1 Lloyd's Rep 692.
91 *In re Wheatcroft* (1877) 6 Ch 97.

The normal absolute minimum period for retaining any files is 7 years. That is because 6 years is the time limit for limitation of action where a simple contract is concerned and allowance must be made for the fact that a Claim Form (which used to be called a Writ) can be held for some months before serving. It is also the time limit for holding on to records for company and Inland Revenue purposes. Therefore, if files are to be destroyed before 7 years has expired, there must be a good, positive reason for doing so. There are some obvious examples. It may be that the files relate to a small kitchen extension job which has itself been demolished to make way for another, larger, extension.

Clearly, if a matter is already in litigation or arbitration, on no account must any files be destroyed. Indeed, a person has a duty to preserve all such relevant files and disclose them to the other party if required.

It is also important to understand from when the 7 years runs. It does not run from the date at which the documents were generated, but usually from the last date when the professional had any dealings with that project. In this context, it should be remembered that the period may be restarted or interrupted if the client, however informally, sought additional advice.[92]

That does not mean that as soon as the 7 years has expired it is safe to dump the files; certainly not. For example, many contracts are executed as deeds and the limitation period is 12 years (say 13 years for safety). Under the Latent Damage Act 1986, which amended the Limitation Act 1980, the long stop period for action in negligence was fixed at 15 years. That may seem an exceptionally long period of time, but even that period may be extended if it does not begin to run until certain other criteria are satisfied. For example, it may be dependent on an indemnity. Fortunately, actions in the tort of negligence against professionals are not particularly easy to mount.

In addition, it is always important to examine very carefully the original terms of engagement for a particular project. If they were drafted by the client's solicitors, they may have extended the limitation period under the contract. This is done by crafty wording more often than may be realised.

Generally, it is inadvisable to dispose of any agreements, letters of intent or legal documents such as leases. The safest thing is to keep them for ever.

92 *Kaliszewska v John Clague & Partners* (1984) 5 Con LR 62.

No documents should be destroyed unless the professional's professional indemnity insurers are happy; perhaps it is going too far to say that they must be happy, but they must not raise any objection.

Some firms have a specific written policy about destroying documents. It can sound impressive when a request for a document is met with the response that 'it is company policy to shred all documents 6 years after their generation'. However, the company will be the loser if a document that could be important in defending the company against legal action has been destroyed. It is no defence at all to quote company policy in these circumstances. The other party will not simply go away.

The question can be answered only on a project-by-project basis, running through a checklist of criteria, including:

- whether it is more than 7 years since there have been any dealings;
- whether a deed is involved that would extend the period to 13 years;
- whether it is likely that any claim in negligence could be brought that would extend the period to at least 15 years;
- whether all fees have been paid;
- whether there is anything about which the professional has always had a nasty feeling;
- whether the professional indemnity insurers have raised any objections.

If the files pass all these tests, the professional must try to think if there are any other reasons for keeping the files. If there are, they must be kept.

Finally, the method of disposal must be carefully considered. Some people favour offering the files to the client. This may be perfectly satisfactory if the files are clean and contain only material that the client has already seen. However, if the files contain letters covered in scribbled notes and comments or if the files have documents that the client has not previously seen, it is best not to go down this avenue.

The only way to get rid of files properly is to physically destroy them by burning (not very eco-friendly) or by shredding. There are many firms that specialise in such work, but it is essential that the firm is reliable and known. Files cannot simply be taken to the nearest wastepaper collection centre, because they inevitably contain confidential information concerning the professional or the client. News reports occasionally highlight instances where personal records of one kind or another are found blowing about the streets. The fallout of such instances so far as the professional is concerned is disastrous.

59 Is an architect liable for the specification of a product which is defective?

Under a traditional contract, the architect is liable for designing the Works and the contractor is responsible for providing the goods and materials and constructing the Works. Specifying is part of the design process. In its broadest sense an architect can be said to be specifying when producing drawings showing what is to be constructed and the constituent elements. Specification notes may be inserted on drawings. Therefore, there is little doubt that specification as a whole is something for which the architect is liable. If something is specified which is defective there is a simple question to be asked: did the architect specify something which was inherently defective or something which was entirely suitable, but of which a defective example was supplied to site?

Clearly, in the second instance, the architect has no liability for the specification. Having specified something appropriate, a defective example was supplied and must be replaced at the contractor's own cost. If the architect specifies something which is inherently defective, the kind of thing which would never be suitable, a further enquiry would be necessary to discover whether the architect had used reasonable skill and care in the specification. If the problem involved the specification of a product for a particular application in the building, the architect would be expected to carefully check all the technical information with assistance from other consultants if the product was specialised, make enquiries of the manufacturers about the products suitability for the application and investigate instances where it had been used previously in similar applications. If, after the use of reasonable skill and care the product was found to be defective, it is unlikely that the architect would be liable and redress might be obtained against the manufacturer, depending on circumstances and, importantly, whether the manufacturer had confirmed the suitability of the product to the architect in writing. Usually, where the failure is in the specification rather than the product, the architect will be liable, perhaps because the specification was insufficiently detailed.[93] It often happens that an architect will specify by reference to an approved sample. What is the position if the sample itself is subsequently found to have the same characteristics as the supplied products which become defective? By specifying the product to be the same as the sample has the architect specified negligently? This question was considered in *Adcock's Trustees v Bridge RDC* where the engineer had approved brick samples for use in inspection chambers.[94] In that case, the bricks supplied were equal to the sample approved and the contractor was not at fault, but the bricks specified and approved were not adequate for the wet conditions actually encountered. However, the court made clear that it was the 'apparent' sample with which the supplied bricks had to conform. If the sample had had concealed cracks, the contractor would not have complied by supplying bricks with concealed cracks which later failed. Bricks without cracks should have been supplied.

60 An architect has been appointed for work on which another architect has been engaged; is there a problem?

There are two major problems which arise when an architect is engaged on a project from which a previous architect has been dismissed or resigned. One concerns the professional obligation not to seek to supplant another architect. The other concerns copyright in any material produced by the previous architect.

The RIBA Code of Professional Conduct does not expressly refer to taking over from a previous architect, but information is contained in the excellent Guidance Notes, particularly notes 4 and 7. In essence, an architect must not actively seek to take on a commission on which another architect has been appointed. However, if approached by a client to take on a commission, an architect should try to establish whether another architect was previously engaged. If there was a previous architect, the new architect must write to the former architect, giving formal notification of appointment. It may be that the previous architect is unhappy, because of the circumstances in which his or her appointment has been terminated.

93 *Rotherham Metropolitan Borough Council v Frank Haslam Milan and Co Ltd* (1996) 59 Con LR 33.
94 (1911) 75 JP 241.

Perhaps there are fees owing or the client may be taking some kind of legal action. Often, on receipt of notification that a new architect is taking the commission, a former architect will write, stating that the new architect should not take the commission and setting out a litany of problems with the client. None of this should concern the new architect unduly. Having notified the previous architect, the new architect can simply get on with the job. Obviously, something written by the previous architect may cause the new architect to think carefully about proceeding, but there is no requirement to refuse to work until the previous architect has been paid in full. That is because it is no part of the architect's duty to pass judgment on whether payment has been withheld rightly or wrongfully; that is a matter for the courts. Ideally, the previous engagement should have been amicably brought to an end and all fees due paid in full, but life is rarely like that.

Copyright tends to be the most serious issue. If the new architect will be using any material produced by the former architect, it is important that the client has a licence allowing such use. The terms of engagement of the previous architect should make the copyright position clear, but disputes over unpaid fees may muddy the situation. Copyright issues are notoriously complex and the architect is not expected to be an intellectual property lawyer. Therefore, if there is the slightest doubt, the architect should obtain an indemnity from the client. This is a document, possibly in letter form, in which the client promises to indemnify the architect against any liability, claims, etc. in connection with infringement of copyright. It is not something to be scribbled hastily by the client in order to get the project underway. The letter should be specially drafted by a lawyer experienced in this field. If satisfied that there is an existing licence which allows the client and any instructed professional to use and change copyright material or having received a satisfactory indemnity from the client, the new architect can proceed with the project.

61 Being asked to use reasonable endeavours sounds less onerous than best endeavours; is that correct?

That is correct. An obligation to use best endeavours has been defined as a duty to leave no stone unturned and to take all reasonable courses of action. It is not an absolute obligation to ensure that something happens and it probably depends on the particular circumstances.[95] Thus an obligation in SBC clause 2.28.6.1 that the contractor will constantly use best endeavours to prevent delay does not amount to a guarantee that there will be no delay. If that was the case, there would be no need for an extension of time clause. In contrast, it has been held that an obligation to use reasonable endeavours only requires a party to take one reasonable course of action even though there may be several different courses available.[96] It follows that if a party is asked to use 'all reasonable endeavours' it is the same as being asked to use best endeavours because both require that all the available reasonable courses of action are taken.

62 An architect's terms of appointment often require co-ordinating, liaising and monitoring. These seem like woolly terms: what do they imply?

They are quite woolly terms which is why they are undesirable in a legal document. A cynic might say that they are often used precisely because they are somewhat woolly; the idea being that a client who requires an architect to co-ordinate, liaise with and monitor other consultants can use the terms to make the architect responsible for any shortcomings of the consultants. Whatever may be the truth behind the inclusion of such terms in appointment documents, architects should not simply sign such appointments, but should seek to have them clarified. Each party to a contract is entitled to know exactly what they are undertaking.

Architects who are in the position of having signed an appointment which includes these terms without clarification are left to determine their ordinary English meanings:

- To co-ordinate is defined as to arrange the elements of a complex whole or to negotiate or to work with others.
- To liaise is defined as to establish a co-operative link or relationship.
- To monitor is defined as to keep watch over, to record and test or to control.

95 *Midland Land Reclamation Ltd & Another v Warren Energy Ltd* (1997) EWHC375 (TCC).

96 *Rhodia International Holdings Ltd and another v Huntsman International LLC* [2007] EWHC 292.

It is suggested that architects agreeing to carry out these duties in respect of other consultants might well be expected, at one extreme, to oversee other consultants on a day-to-day basis and to check and control everything they do. This would give the architects in question an unreasonable amount of responsibility for the services of consultants operating in disciplines of which the average architect may have but slight or merely general knowledge. Few clients would actually expect that degree of control by their architects and few consultants would be happy with the situation even if there were any architects capable or exercising such close supervision.

A great danger with woolly terms such as these is that the architect will think they simply refer to the kind of relationship between architect, as lead consultant, and other consultants as the architect has come, by experience, to believe is normal. Consultants will have a similar view, but the precise details of such a relationship will probably vary from the architect's experience (unless they frequently work together) while the client, who may be building for the first time, may have a completely different idea based on the definitions noted above. The questions, of course, do not arise unless problems appear on site.

63 What are the implications of an architect being asked to provide a certificate of readiness for the preparation of bills of quantities?

A certificate of readiness is properly the name given to a document which may be required by the court from a party to proceedings in that court to indicate that everything necessary has been done to allow proceedings to commence. The term is not one which is generally used in the construction industry. From the context, it seems that the architect in question is being asked by the quantity surveyor to certify that everything is ready for commencement of the preparation of the bills of quantities. It is important to understand that, when asked to certify, the architect is being asked to form a professional opinion and to express that opinion in a formal way. It is one thing for a quantity surveyor to ask the architect if all the information is ready for bill preparation to start, but it is quite another to ask for a certificate to that effect. It implies that the architect has given very serious and in-depth consideration to the matter and that his or her professional reputation may be at stake.

On the one hand, it is understandable that a quantity surveyor does not want to embark upon the preparation of bills only to discover that there are many revisions to the information which has been supplied. Obviously, the quantity surveyor would prefer if the whole scheme was entirely frozen to allow the bills to be completed. No doubt some architects are guilty of providing some information to the quantity surveyor and then following it with a stream of further information and changes. This can be very irritating. It is no doubt irritating to architects when clients ask for changes long after what should have been the last minute. The fact is that the nature of the design, tendering and construction process is such that for any of the professionals involved to say that the documents are fixed and that no further changes will be made would be foolhardy to say the least. Therefore, an architect asked for a certificate of readiness should refuse. The quantity surveyor is entitled to know that the bulk of what has been supplied is unlikely to change, but it can never be guaranteed, certainly not by the architect who must take instructions from the client. For a quantity surveyor to take an inflexible line and to expect the architect to guarantee something over which the architect does not have final control is completely unrealistic.

64 Are there any dangers for the architect if the client wants full drawings, but intends to deal directly with a contractor to get the project built?

It is by no means rare for an architect to be asked to carry out a full service for a client, but for the client to decide to proceed without the architect for the construction stage of the project. Sometimes, a client will engage an architect only for the production of construction drawings and the obtaining of all necessary statutory permissions. One assumes that the client is motivated by a desire to save the costs of professional fees for contract administration. Clients may sometimes try to get the architect to agree to be available to visit site or simply answer questions over the telephone as and when required by the client or by the contractor. Architects are naturally concerned about their liabilities, for example, if there are any defects in the drawings which no one notices until the project is complete. Is it not the case that the client has paid for a set of drawings which should be perfect and that any errors on drawings leading to defective construction are entirely the architect's responsibility?

Like the answers to most general questions, the answer to this one is that it all depends on the particular circumstances. A good rule of thumb for architects is that they should either accept commissions for a full service or for a partial service, but never under any circumstances for a mixture of the two. Where an architect carries out a full service in connection with a traditional contract it is clear that the architect is liable for any failure in the design or any negligent advice given during the progress of the Works.

Where the architect is acting as contract administrator and regularly inspecting the work as it progresses, it is likely that any discrepancies or errors in the drawings will be picked up and corrected as the work progresses. If they are not discovered until after construction has taken place, the architect may be liable for the consequences after taking account of whether the contractor had a duty to warn of the error in the particular circumstances. Some discrepancies are virtually inevitable in any set of drawings of any complexity and, although the architect has a duty to provide correct information to the contractor, errors are something which are generally recognised and dealt with in practice as the work proceeds.

If the architect is employed only up to tender stage and, at the client's request, ceases work at that point and takes no further part in the project, neither visiting the site nor answering questions over the telephone, it is likely that the architect will have no liability for the consequences of errors on the drawings which lead to defects in construction. That is because, after tender stage, the client has ceased to rely on the architect and from then onwards relies on the expertise of the contractor.[97] Whether and to what extent the contractor may be liable for failure to discover and deal with defects on the drawings will depend on the extent to which the contractor is aware and accepts that it is relied upon by the client. The danger for the architect lies in becoming involved in the project during the construction stage. If the architect agrees to answer any questions or to visit site to deal with any problem, it will be difficult to contend at a later date that the architect was not aware of any other problems on the site. Although the architect may have terms of engagement which exclude any involvement during construction, that architect probably will be assuming responsibility by accepting a subsequent invitation to visit site or to become involved by giving further advice. As a professional

97 *Brunswick Construction Ltd v Nowlan* (1974) 21 BLR 27.

person, it is for the architect to advise the client about the degree of professional inspection and involvement required. It is not for the architect to dictate such matters. The client's option is to accept the services which the architect advises are necessary or not to do so. Would solicitors agree to produce all the documents required for a trial and subsequently only attend court as and when the client decided, leaving aside that the court rules would not allow it? Would a surgeon perform a serious operation and then agree to see the patient for follow up consultations in accordance with the patient's wishes? One only has to think of comparable situations in other professions to see that to expect the architect to attend site only as required by the employer or the contractor is nonsensical.

Chapter 9

Fees

65 If the tender price has been reduced and the architect has been paid for the reduction work, is the architect entitled to be paid for doing the extra design in the first place?

Unfortunately it is only too common for the lowest tender to be returned in excess of the client's budget figure. This can be due to several causes, only one of which need seriously concern the architect.

If the architect has satisfied the client's brief in terms of design and price, the architect is entitled to the proper fee to the particular stage. It may be argued that, if a tender is returned higher than the figure the client wishes to pay, the architect has not satisfied the brief and, therefore, has no entitlement to the full fee. There are two main reasons why a tender may be high. The first, beloved of architects and quantity surveyors alike but mystifying to clients, is that the general climate of tendering is high. There are variations on this theme, such as that the climate is high in that particular area or that there is a lot of work about and none of the good builders is interested in taking on these Works. None of these explanations or excuses really accounts for the fact that the architect or quantity surveyor ought to be well aware of current trends in tendering. It is rare that the 'general climate' excuse is a good reason for a larger than expected price unless the change in climate is quite sudden.

The other main reason is that a mistake has been made in the costing of the job. If the architect did the costing, however informally, that is a problem. The architect should not have a problem if the costing was done by the client's directly appointed quantity surveyor, unless the quantity surveyor can show either that the architect did not provide enough information for the cost estimates, or that the architect did not follow the cost plan.

There are other reasons of course; relatively common is the client who deliberately increases the cost of a project despite being warned of the consequences. The architect should always keep a careful record of such instructions and take the precaution of putting advice about the burgeoning cost in writing.

If a high tender has been returned, but the client has agreed to pay the architect's fees to carry out reductions to the drawings and specification, it suggests that the client is not pointing any finger of blame against the architect. However, if the blame lay with the architect, clearly the client ought not to be required to pay fees to the architect to rectify the error. But in such circumstances, clients may abandon logic and argue that they are not prepared to pay fees for the design of just those parts of the project that the architect is so assiduously removing so as to lower the cost. Again, if the architect was culpable, the client would be correct and no fees would be payable for the design or the removal of the costly parts. However, it follows that an architect who is not culpable is entitled to both sets of fees, just as an architect who is culpable is entitled to none.

Where an architect culpably fails to design a building to the client's budget the architect cannot claim fees for work aimed at reducing the tenders. Moreover, the architect is not entitled to fees for that part of the original design which is over budget. That may be difficult to determine accurately and the deduction of a percentage is a simple solution.

66 Can the architect claim extra fees for looking at claims?

This question usually arises when the architect has been engaged on the basis of SFA/99 or CE/99 (April 2004 updates), S-10-A or C-10-A which are in somewhat similar terms. In order to arrive at the answer to this question, it is necessary to look at the terms of the engagement. A client will usually argue that dealing with contractors' claims is part of the architect's normal contract administration duties. However, if the architect is being remunerated on a percentage basis (as is usual), the client will also argue that there is no incentive for the architect to reduce a claim made by the contractor, because the architect's percentage payment will increase as the claim increases. Such an argument says little for the client's opinion of the architect's professional integrity and it is a source of wonder that a client will employ an architect while having such reservations. SFA/99, CE/99 and S-10-A lay that particular concern to rest, because under 'Definitions' it clearly states that the Construction Cost does not include 'any loss and/or expense payments paid to a contractor . . .'. A similar term would be implied in C-10-A.

For a fully designed project the architect's management services include 'Administering the building contract'. Although a number of activities are stated as being included, dealing with the contractor's claims is not one of them. Indeed, dealing with contractor's claims is one of the additional services which must be specifically added to the basic services if required. Each of the appointment documents contains provision in greater or less detail for additional fees for dealing with claims. SFA/99, CE/99 and S-10-A expressly refer to delay, disruption and prolongation, while C-10-A refers to 'reasons beyond the Architect's control', which must include contractor's claims. It is to be noted that the phrase 'contractor's claims' is broad enough to cover extensions of time and additional preliminaries in relation to variations as well as loss and/or expense.

Therefore, the architect is entitled to claim additional fees on a time basis for dealing with claims of all kinds provided that the claim has not arisen as a result of a breach of the engagement by the architect. Architects claiming additional fees are well advised to notify the client as soon as it becomes clear that extra fees will be involved and detailed timesheets should be kept setting out exactly what the architect does in relation to the claim. The timesheets of many architects are totally inadequate as a record of what was actually done.

67 If the architect charges additional fees for work which has to be done as a result of the contractor's breach of some obligation, can the employer recover such fees from the contractor?

There are two important points to consider. The first is that, if the contractor is in breach of an obligation under the contract, the general law allows the employer to recover damages for the breach. The measure of such damages is that the employer is entitled to be put back into the position he or she would have been in, so far as money can do that, as if the contractor had properly performed its obligations.[98] The second point is that, under RIBA appointment documents, the architect is entitled to claim additional fees if engaged in additional work due to matters beyond the architect's control.

Therefore, if it can be established that the contractor is in breach of contract and if the breach is the direct cause of additional work being required from the architect, the architect's fees associated with the additional work would be the damage suffered by the employer as a result of the breach. Therefore, it follows that the employer is entitled to recover the amount of such fees from the contractor. The easiest way for the employer to do that is to issue a withholding notice for the amount and deduct it from payment due to the contractor under a certificate. If no further interim certificates are due, the employer may either issue the withholding notice for the final certificate and deduct the amount from the final payment or take legal action to recover damages for breach of contract. (Note that Part 8 of the Local Democracy, Economic Development and Construction Act 2009 will amend the notice provisions, but at the time of writing, no date has been set for commencement.) Adjudication may be the simplest route, but more complex matters might better be dealt with in arbitration or litigation as permitted under the terms of the contract.

An overriding consideration is whether the building contract contains machinery for dealing with the breach in question. If so, it is that machinery which will take precedence over the common law position. For example, if the contractor is in breach of its obligation to complete the Works by the completion date in the contract, it is a breach of contract for which the employer is entitled to damages. However, under most standard form building contracts there is provision for liquidated damages in such circumstances which will determine what remedy is available to the employer. All the costs resulting from late completion, including additional professional fees, are deemed to be included. Therefore, the employer would not be able to make a separate claim for architect's fees. There is detailed provision in most building contracts for remedies if the contractor fails to construct the Works in accordance with the contract. Normally, it is those remedies which will apply. Sometimes, clauses are made subject to any other rights and remedies of the parties. SBC

98 *Robinson v Harman* (1848) 1 Ex 850.

clause 8.3.1 is an example. In that case, the parties are not restricted to the remedies set out in the contract and may seek common law remedies instead or in addition. Obviously, the employer cannot recover twice for the same breach.

Less clear is the situation where the architect is entitled to charge additional fees, but it may not be clear that the contractor is in breach. For example, it is not unknown for a contractor to notify the architect that practical completion has been achieved, but the architect finds, on inspection, that is not the case. Sometimes such notifications and abortive inspections occur several times. It is perhaps open to debate whether the contractor's notification, giving incorrect information, is a breach of contract. On balance, it possibly is in breach of an implied term that the contractor should co-operate with the employer to execute the Works. If the preliminaries to the bills of quantities or the specification expressly require the contractor to notify the architect when practical completion has been achieved, a failure to do so properly, that is to say prematurely, is certainly a breach. What of contractors who submit applications for loss and/or expense under clause 4.23 of SBC which are twice or three times the amount to which they are entitled? There is probably a valid case to be made that a contractor who submits a grossly inflated claim, and thereby involves the architect in substantial expenditure of time and resources in dealing with it, is in breach of its obligation under clause 4.23. The argument would probably be that the contractor is entitled to submit an application for loss and/or expense as permitted under the contract, but plainly it is not entitled to submit a claim for loss and/or expense which is in excess of what is allowed under the contract. Indeed, a contractor which makes a claim for money which is demonstrably wrong may have serious questions to answer related to obtaining money by deception.

68 Can the architect claim extra fees if another contractor is engaged after termination?

The important question is whether the termination has taken place as a result of some default on the part of the architect. If not, the engagement of another contractor to complete the Works following termination is certainly additional work beyond the architect's control and, therefore, the subject of additional fees. The calculation of the additional fees will require some care. Take the situation where the termination has taken place and the Works are only about 50 percent complete. The architect will be involved in organising tenders for completion and a survey of the completed work may be necessary. It is possible that a new specification must be drafted and there will be a lot of old ground to cover again with the new contractor. However, it is not sufficient for the architect simply to log all the time expended from the notice of termination until practical completion under the new contract, because the architect would have been involved in the final 50 percent of the Works in any event and that cannot be additional work although there may be some aspects which are additional.

Where there has been a default on the part of the architect which has led to termination by the contractor, the architect, if not already dismissed by the employer, would be extremely unwise to attempt to charge for additional fees in dealing with completion with another contractor. If the employer can show that the termination is a direct result of the architect's default, that would found an action for damages against the architect for all losses suffered by the employer as a result of the termination. Even if the architect had the nerve to submit a fee account for the additional work, the amount claimed would simply rank as damages for which the employer could claim from the architect. Under the RIBA Standard Agreement (S-10-A) clause 5.9 the clause is expressly stated not to apply where there is a breach of the agreement by the architect. Obviously, any negligent act of the architect which resulted in the contractor successfully terminating its employment would be a breach of the architect's obligation under the agreement to use reasonable skill and care (clause 2.1). However, examination of the grounds for contractor termination in SBC (clauses 8.9 and 8.11) suggests that such termination as a result of the architect's default will be relatively rare.

69 If the architect was engaged on the basis of an RIBA Agreement and the client is being funded by a public body such as the Lottery, must the architect wait for fees until funding comes through?

Although one is told never to generalise, it is difficult not to do so and, therefore, one might say that architects, as a class, are not noted for their keen business sense, particularly where their own contracts with their clients are concerned. This is something of which many clients are not slow to take advantage.

Architects enter into contracts with their clients in many ways. Although both the RIBA and the ARB Codes of Conduct require architects to consign all their appointments to writing, setting out key points regarding fees, services to be provided and so on, it is still the case that some architects find themselves working for clients on contracts which, if they exist at all, are purely oral. If the clients of these architects drag their heels on payment, there is little to be done without complex legal action.

The remainder of architects either contract on the basis of an exchange of letters, one of the standard RIBA forms or a bespoke set of terms drafted for the client by solicitors – in other words, in writing. That is very important, because agreements in writing fall under the Housing Grants, Construction and Regeneration Act 1996.[99] The Act does not apply to ordinary consumers constructing residential property for themselves, but it is a very useful Act so far as other clients are concerned. At the time of writing, Part 8 of the Local Democracy, Economic Development and Construction Act 2009 is not yet in force. When it is, among other things, it will amend the 1996 Act to remove the need for construction contracts to be in writing before the Act applies to them.

SFA/99, S-10-A or similar RIBA terms comply with the current Act. Therefore, they set out the way in which payment must be made and what is to happen if the client wishes to withhold payment. They also include provision for adjudication – a quick and relatively inexpensive way of settling disputes.

It is quite common for a client to be reliant upon funding from elsewhere. Typically this might be a bank, a mortgage provider, an insurance company or some body such as the Lottery. Perhaps the appointment document includes a term which says that, notwithstanding the payment provisions (which are probably for monthly payment), the architect will be entitled to payment only when funding comes through to enable the client to pay. Sometimes, there is nothing in the terms and the client simply springs the surprise on the architect at their first design meeting or, more likely, not until a fee account becomes seriously overdue. Many architects try their best to

99 It has already been noted that Part 8 of the Local Democracy, Economic Development and Construction Act 2009 will amend the 1996 Act so as to remove the stipulation that provisions of the Act apply only to construction contracts in writing. At the time of writing, the commencement date for Part 8 had not been fixed.

accommodate such clients and do not press for payment until funding appears. They do this partly because there is no point in pressing someone who has no money for payment, and partly because architects naturally try to assist their clients, usually far beyond what is required by the law, codes of conduct or sometimes even common sense.

It is surprising that a client will even consider embarking on expensive construction work without having the necessary means to pay and trusting that money will be made available in time. There is no doubt that, under RIBA Agreements, the client is obliged to pay the architect the amount stated on the invoice if no withholding notice has been served.

The architect is not simply entitled to stop work if payment is not made. However, the architect is entitled, under RIBA Agreements and under section 112 of the Act, to suspend performance of all obligations if payment is overdue provided that at least 7 days' notice in writing is given stating the intention to suspend and the grounds for doing so. Few architects seem to avail themselves of this right, although nothing concentrates a client's mind so well as the knowledge that all work will stop in 7 days. The right is valuable because, contrary to popular belief, there is no such right of suspension under the general law if the client fails to pay, although consistent failure to pay might be grounds for treating the client's defaults as repudiation.

There is another important part of the Act that is not reflected in RIBA Agreements nor in most other construction contracts. This is a provision that strikes at the heart of the 'pay when paid' ethos. The key part is as follows:

> Section 113(1) A provision making payment under a construction contract conditional on the payer receiving payment from a third person is ineffective, unless that third person, or any other person payment by whom is under the contract (directly or indirectly) a condition of payment by that third person, is insolvent.

Therefore, if a client attempts to insert a clause making payment dependent on receiving funds from elsewhere, the clause is ineffective unless the supplier of the funds becomes insolvent. The likelihood of a bank, building society or the Lottery becoming insolvent, while not negligible, is not significant. Therefore, there is no need for the architect to act as the source of a bridging loan until money comes through: it is up to the client to make its own arrangements.

Where the client simply notifies the architect that payment is dependent on a third party after the terms of engagement have been agreed, the architect can, and usually should, simply reject the statement out of hand. There is a danger if the architect gives the client to understand that the architect will wait for the money until it arrives from the funder. In those circumstances, the architect may be estopped (prevented) from subsequently having a change of mind and demanding the money. Estoppel is a legal principle with many facets. In this instance, the principle is that, if the architect represents to the client that strict legal rights will not be enforced and if the client has acted to its detriment in reliance on that representation (such as in this case using its own money which might have been used to settle the architect's debt), the architect is prevented from going back on the representation.

The moral is that an architect should insist on being paid on a regular basis. If the architect is moved to allow the client time to pay, that time should have a defined end, which must be clearly set out in writing and acknowledged by the client. Far better for the architect to use the tools set out in the contract and the Act: the right to payment, suspension on failure to pay, adjudication, arbitration or litigation appropriate to secure payment and statutory interest on late payment.

70 The architect agreed a fee of 5 percent of the total construction cost. The contract sum was £325,000, but it is now only £185,000 at final account stage. Is the architect obliged to return some fees?

Many quantity surveyors work on the basis that if they inform the client throughout the contract that the final account is likely to be £x and if, at the end of the project, they are able to tell the client that the final account figure is actually £x minus £20,000, the client will be surprised and grateful and consider the project a huge success. There is a great danger in this approach, which is that the client is anything but grateful and considers that if it had not been for the incompetent reporting of the projected final account throughout the progress of the Works, the client would have known that there was this money to spare, which could have been readily spent on upgrading an important part of the building. The approach smacks of the surgeon informing the patient that death has been avoided when the patient actually would like to be made well.

For architects, this kind of last-minute revelation can be catastrophic. That is because it is common for architects to base their percentage fee instalments on the latest valuation or final account forecast by the quantity surveyor. If the quantity surveyor quite suddenly at the end of the project values the final account at something substantially less than forecast, the architect is left with the prospect of having to reimburse some fees.

In this question, the architect is entitled to 5 per cent of £185,000. However, there is a big question to be asked about how the contract sum dropped so markedly. If it was as a result of the client requesting savings, the architect is of course entitled, certainly under the RIBA Agreements, to recover the cost of redrawing or re-specifying, probably at an hourly rate. Where RIBA Agreements have not been used, the position will depend on the agreement between architect and client, but in most instances the architect should be able to charge for the extra work provided due notice is given to the client before the extra work is undertaken. If the quantity surveyor properly forecast the downward trend in the final account as savings were made, the architect ought to have been aware of the likely outcome and made provision accordingly. Of course, if the quantity surveyor maintained a high forecast until the last minute before dropping it, it may amount to professional negligence.

The answer to the question of what the architect can recover in fees depends on the terms of engagement. If they say 5 per cent of the total construction cost, that is what the architect can charge. If the architect has inadvertently charged more, there must be a reimbursement. If the architect has done extra work, there can be an additional charge, usually on a time basis. There are many pitfalls in claiming fees, often because the parties have not looked far enough ahead or thought of all the possibilities before signing a contract.

Many architects think that to base a percentage fee on the lowest acceptable tender will save the day if the client decides not to proceed at tender stage. Such architects commonly confuse the lowest *acceptable* tender with the lowest tender. If the client does not consider any tender is acceptable, the architect may have a difficult task in arriving at a figure on which to base the percentage fee.

71 Can the architect recover interest on unpaid invoices if the client can show that they are incorrect?

Before 1998, there was no automatic interest on debts. In order to decide whether interest was payable, one had to look at the contract provisions or whether or not interest could be claimed as special damages. All that changed in 1998 when the Late Payment of Commercial Debts (Interest) Act 1998 started to come into force. It was not fully in force until 2002, because the kinds of organisations who could claim under the Act were progressively increased during that period. The current position is that if an invoice from one commercial organisation to another becomes overdue, interest is claimable from the date it became overdue at the rate of 8 percent over the Bank of England Base Rate calculated for 6-monthly periods from the 30 June or the 31 December. That is quite a high interest rate even if the Base Rate is low. In addition, the party claiming interest can also claim a lump sum payment for each unpaid invoice ranging from £40 to £100 depending on the size of the invoice. The Act may not apply if the contract provides for a substantial remedy. No interest or lump sum at all can be claimed if it can be shown that the money claimed on the invoice is not actually due.

The answer to the question is to be found in a recent case.[100] A plant hire company had submitted invoices, for cleansing and decontamination work in connection with an outbreak of swine fever and foot and mouth disease, which were disputed. In fact, many of them were found to be wrong in that in various ways they were claiming too much. The Secretary of State argued that the Act did not apply until he was in possession of completely accurate invoices and they had become overdue. The matter went to the Court of Appeal.

In a judgment which is of interest to all construction professionals and others who are seeking payment in a difficult financial climate, the court drew attention to section 4(5) of the Act which states that the 'relevant day' (the last date for payment) is calculated from the date on which the obligation for which the debt arises is performed or the date on which the purchaser has notice of the amount of the debt 'or the sum which the supplier claims is the amount of the debt', whichever is later. The court held that section, 4(5) anticipated that

100 *Ruttle Plant Hire Ltd v Secretary of State for Environment, Food and Rural Affairs* [2009] EWCA Civ 97.

the supplier may take a provisional view of the amount due which would be subject to review, but that did not prevent interest from accruing on whatever was eventually found to be the correct amount. Importantly, the court held that the Act, as a matter of policy, did not require an invoice to be perfect. If it did, a purchaser would simply have to find the smallest error in order to avoid paying interest. It was noted that the invoices had been accompanied by supporting information such as timesheets which made it possible for the Secretary of State to calculate the actual amount due and to pay it.

It is a sad fact that many people and organisations make no attempt to pay invoices until under threat of legal proceedings and, at that point, they reveal some error in an invoice hoping to cause the creditor to have to start again and re-submit another invoice. This case shows that delaying payment on that ground will not work and that interest on the amount properly due will still accrue and be payable. The Act's purpose was to encourage prompt payment. Late payment is the biggest evil in the construction industry. If all parties paid debts within 7 days, cash flow problems would be largely eradicated. It is the unavailability of money to pay debts which causes businesses to fail even though on paper they may be in substantial profit. All parties, construction professionals, contractors, sub-contractors and suppliers should not hesitate to get copies of the Act and apply it as soon as invoices become overdue. A very useful website is www.payontime.co.uk.

Chapter 10

Design

72 Can the architect escape liability for defective design by delegating it to a sub-contractor?

In general, the answer to this question is 'No'. However, the precise terms of the architect's engagement by a client will obviously affect that statement. It is possible that an architect can have exclusions of liability for design written into the terms of engagement with a commercial client, but note that such terms will be effective only if they satisfy the test of reasonableness in the Unfair Contract Terms Act 1977. If the architect has entered into an engagement with a client on one of the RIBA Agreements or on a simple exchange of letters, that architect will have overall responsibility for design of the project. The only way in which the architect can avoid that liability is if the client specifically so agrees.

In *Moresk Cleaners Ltd v Thomas Henwood Hicks*,[101] the client was a dry-cleaning and laundry company. It engaged the architect to prepare plans and specifications for the extension of its laundry. Unknown to the client, the architect had delegated the design of part of the building to specialist sub-contractors. Subsequently, cracks appeared in the structure that were found to be design defects. It was held that the architect had no power to delegate duties to others without the permission of the client.

The RIBA Agreements contain important provisions that allow the architect to advise on the need to appoint consultants to carry out specialist design. In such instances, the architect is clearly not liable for any defects in the consultants' designs. This point is emphasised in SFA/99, CE/99, S-10-A and C-10-A where the client undertakes

101 (1966) 4 BLR 50.

that where work or services are performed by any person other than the architect, the client will hold that person responsible for the competence and performance of the services and for visits to site. This is a most important provision that protects the architect, and a forerunner of this clause has been upheld by the courts for the architect's benefit.

Obviously, if the architect fails to advise a client on the appointment of specialist designers, the client will be entitled to assume that the architect will retain design responsibility in those areas. It is not sufficient that the specification or bills of quantities refer to design by specialist consultants or sub-contractors, nor even that the contract versions incorporating designed portions are used (such as in ICD or MWD). The architect will be responsible to the client for all design unless the client has been expressly informed otherwise and has consented. Clearly this is best done in writing. An architect who does advise the transfer of design responsibility to others has a clear-cut duty to ensure that appropriate contracts and warranties are put in place to protect the client in the event that there are defects in the transferred designs. Failure to put such matters in train may well amount to serious professional negligence on the part of the architect.

73 Is the contractor liable for something done on its own initiative?

It is surprising how often this question arises. It arises when, typically, a contractor, working under a JCT traditional contract, knows broadly what is required, but does not have drawings or specification which shows precisely what is required. The contractor thinks it knows what to do and carries on with the work to its own detail. Subsequently, the detail fails with serious consequences. It might be a badly constructed roof detail, an inadequate stanchion base or perhaps wrongly positioned heating pipes. Who is liable? Under SBC, IC and MW the contractor has no liability for design. Even where ICD and MWD are used, the contractor's liability for design is confined to those items clearly listed in the contract particulars. Under a traditional contract, the architect designs and the contractor constructs in accordance with the designs.

Provided that the contractor constructs exactly as drawn or specified, it is likely to have little liability for the result unless the defect is so obvious that the contractor should have warned the employer of the potential danger. However, the position changes if the contractor is not provided with sufficient information to enable it to construct the particular detail. In such situations, it is for the contractor to ask the architect for the missing information. In practice, a contractor may find itself in a situation where it is keen to make progress, but there is no detail. If it pauses until the detail has been requested and eventually provided, there may be a delay of which the contractor has to notify the architect and hope for an extension of time which is by no means certain. If the contractor, thinking it knows what is required, presses on with the work, it places itself in a difficult situation.

There are two problems for the contractor. Firstly, the architect, on seeing the solution which the contractor has adopted, may instruct the contractor that it is not in accordance with the contract and must be removed. That would be strictly correct. The contractor's work is not in accordance with the contract, because the contract documents do not show what the contractor has done, in fact they show nothing. Secondly, even if the architect does not notice what has been done, it may be defective and cause a problem – in the worst case perhaps a collapse. The contractor cannot blame the architect, because the architect has not designed the detail which has been built. In such circumstances, it is likely that the contractor has assumed responsibility for the design.[102] Therefore, the contractor will probably be liable for the consequences of the defective design provided that they were within the contemplation of the parties as likely to occur at the date the contract was executed. There are probably many such instances, where the contractor unthinkingly assumes design responsibility, but where there are no untoward consequences because the contractor has produced a satisfactory solution. However, the author recently heard of a contractor which became responsible for an entire defective heating system in a building, because it had only been provided with an outline concept drawing showing pipework and equipment in diagrammatic fashion and an outline specification which required the contractor to complete the whole of the design. Instead of informing the architect that it had no design responsibility under the contract, the contractor proceeded to design the whole system including all pipe runs and specified equipment where not otherwise stated. The system was defective and the cost of rectification was claimed from the contractor.

102 *CGA Brown v Carr & Another* [2006] EWCA Civ 785.

74 If the contractor is to 'complete the design', does that mean that existing design can be assumed to be correct?

The answer to that question is that it all depends on the type of contract being used. Under the JCT WCD contract the contractor was required to complete the design of the Works. It used to be thought that the contractor was not being given responsibility for the design as a whole, but merely to complete what was, presumably, left incomplete. That seems the sensible view. However, that view changed with the judgment in *Co-operative Insurance Society Ltd v Henry Boot Scotland Ltd* (2002).[103] Although this case concerned the JCT 80 form of contract as amended by the Contractor's Designed Portion Supplement, the CDP had many similarities to the design and build contract, importantly including an obligation that the contractor must 'complete the design for the Contractor's Designed Portion'. Some other clauses were almost identical to the equivalent clauses in the design and build contract. The court took a very clear approach:

> In my judgment the . . . process of completing the design must, it seems to me, involve examining the design at the point at which responsibility is taken over, assessing the assumptions upon which it is based and forming an opinion whether those assumptions are appropriate. Ultimately, in my view, someone who undertakes, on terms such as those of the Contract . . . an obligation to complete a design begun by someone else agrees that the result, however much of the design work was done before the process of completion commenced, will have been prepared with reasonable skill and care. The concept of *'completion'* of a design of necessity, in my judgment, involves the need to understand the principles underlying the work done thus far and to form a view as to its sufficiency. . . . If and insofar as the walls remained incomplete at the date of the Contract, Boot assumed a contractual obligation to complete it, quite apart from any question of producing working drawings.

This judgment makes clear that the contractor's obligation to complete the design under the CDP supplement extends to an obligation to check the design which has already been prepared to make

103 (2002) 84 Con LR 164.

sure that it works. The contractor cannot simply assume that information including design already provided is correct. Because of the similar wording, it appears that the judgment applies to WCD 98 also. Indeed, it is probably not going too far to say that wherever a contract calls upon the contractor to complete the design, its obligation extends to checking the original design before proceeding.

When drafting the new design and build form, DB, the JCT have taken account of this judgment. A new clause 2.11 has been included which provides that the contractor is not responsible for the Employer's Requirements nor for verifying the adequacy of any design contained in them. It seems to follow that if the employer has caused the whole of the design to be prepared by an independent architect and included within the Employer's Requirements, there will be no design to complete and the contractor will have no obligation in this regard. At the other end of the scale, if there is no design included in the Employer's Requirements, the contractor will be responsible for all the design. If, under DB, the contractor is handed a partly finished design, it appears that the contractor's obligation will simply be to complete the design without carrying out any checks, unless of course, there are obvious problems with the original design. This effectively reinstates the position as most people thought was the case until the *Co-operative Insurance* case.

75 Is the contractor liable for design produced by a nominated sub-contractor?

Although the JCT Standard Building Contract no longer has provision for nominated sub-contractors, there are other contracts which do make such provisions, among them the GC/Works/1 contract. There are also many employers who continue to use the 1998 edition of the JCT Standard Form (JCT 98) and many other bespoke contracts which provide for nomination. Therefore, this is still an important question. Clause 35.21 of JCT 98 expressly stated that the contractor was not liable for any design carried out by the nominated sub-contractor, but what is the position where clause 35.21 has been struck out or in another situation where there is no such clause?

The position was considered in *Sinclair and Another v Woods of Winchester Ltd (No 2)*.[104] This was a case where the employer,

104 (2006) 109 Con LR 14.

Mr and Mrs Sinclair, contracted to have a swimming pool constructed by Woods. The matter had gone to arbitration and this was an appeal against the arbitrator's award. In the event, the court refused permission to appeal, but in the course of the trial, the court usefully set out the law on this point:

> Where an employer nominates specialist sub-contractor to carry out work, it will often be because that sub-contractor will be performing a specialist design function, in addition to the actual carrying out of the works on site. In those circumstances: (a) the design work performed by the specialist sub-contractor ought to be the subject of a direct warranty from the specialist sub-contractor to the employer; (b) the carrying out of the work on site may be sub-contracted by the main contractor to the nominated sub-contractor, but the extent to which the main contractor is liable even for defects in the workmanship of the nominated sub-contractor will depend on the precise terms of the various contracts . . .
>
> . . . as far as I am aware, there is no reported case in which it has been held that a main contractor, whose workscope excluded any design, somehow acquired a design liability simply because it entered into a sub-contract with a nominated sub-contractor who was in fact carrying out design work. In my judgment such a finding would be contrary to common sense.

The overall responsibility for design remains with the architect unless the employer has given authorisation for it to be transferred to the nominated sub-contractor. If that is the case, the architect should ensure that there is a suitable warranty in place between employer and nominated sub-contractor to deal with any design defects. In any event, unless expressly stated in the main contract, the contractor has no express design responsibility.

76 If the architect's design is faulty, but the contractor builds it badly, who is liable?

This is the kind of question which crops up time and again, usually when an architect instructs a contractor to rectify defective work and the contractor argues that it is the design which is faulty. The contractor may argue that it matters not at all that it constructed the work badly because faulty design would inevitably result in failure.

This was considered in *Sinclair and Another v Woods of Winchester (No 2)*.[105] This was an appeal from the award of an arbitrator and the court decided that the arbitrator's award was not to be overturned on this point, because among other things the arbitrator was obviously right. In this case, the design was the responsibility of a nominated sub-contractor, but the principle is the same as if it had been an architect's design.

This question has to be answered by deciding which is the underlying or operative cause of the defect. In this case, the arbitrator found that the operative cause was the design and that the errors attributable to the contractor simply compounded the speed of the failures. The answer to the question can only be discovered by a logical analysis of the facts and, therefore, no all-embracing formula can be applied. It is tentatively suggested that if it is established that both design and workmanship are defective, the next question to ask is whether good workmanship would have made any difference. If not, it is likely that the liability will fall entirely on the designer. However, if the defective workmanship caused the defect to appear earlier or to be greater in extent than would have been the case, the liability may be shared. Plainly, there may be instances where the effects of a less than ideal design would be entirely ameliorated by good workmanship and, in such a case, the cause of the defect might well be held to be the poor workmanship.

77 Does the architect have a duty to continue checking the design after the building is complete?

A problem that frequently arises is the architect's duty to review the design: in other words, the architect's duty to check whether the design is, in fact, sufficient. Is there such a duty? And if there is, when does it end? If problems arise after the architect has finished any involvement with the building, must the architect return at the request of the client to check the design?

These questions were considered by the court in the case of *New Islington and Hackney Housing Association Ltd v Pollard Thomas and Edwards Ltd*.[106] Essentially, what the judge was trying to do was decide whether the limitation period had expired so as to protect the

105 (2006) 109 Con LR 14.
106 [2001] BLR 74.

architects against an action for breach of contract and/or negligence. In this instance, the judge did indeed find in favour of the architects.

In deciding whether the architect is under a continuing duty to review the design, the starting point has to be the terms of engagement. In this case, the terms included much what one might expect including 'completing detailed design', and the various parts of stage H: tender action to completion. That included preparing the contract, issuing certificates and performing other administrative tasks and accepting the building for the client. The terms did not expressly include a duty to keep the design under review, still less the duty to keep the design under review after practical completion.

The building contract was IFC 84 and the judge was particularly impressed by the fact that, although the contract allowed the architect to issue instructions requiring variations up to practical completion, the contract did not allow the issue of such instructions after practical completion (the same principle applies to the current traditional JCT contracts). So, although the architect could have altered the design before practical completion, the architect was unable to do so afterwards.

The judge went further. He said that if the client asked the architect to investigate a potential design defect after practical completion, the architect was entitled to refuse or to agree only if the client was prepared to pay a fee – because such work was not part of the original terms of engagement. In any event, even if the architect had a duty to review the design after practical completion, it would arise only if something happened that gave the architect reason to think that a reasonably competent architect ought to review the design.

78 Who owns copyright – client or architect?

The straight answer to this question is that copyright in an architect's design is owned by ('vested in' is the legal phrase) the architect who produced the design. Clients sometimes begin to claim copyright if they fall out with their architects and are disinclined to pay the proper fee. Copyright is governed by the Copyright, Designs and Patents Act 1988 as amended and there is also a substantial amount of case law on the topic. It is quite complex and, as with all the other questions in this book, specific problems require specific answers; therefore all that can be done here is to set out a few general principles.

It is important to understand that copyright does not subsist in ideas, but only in the way in which the ideas are presented. Clients often think that they are just as responsible for the finished design as the architect concerned. In a way that is true. Architects and clients usually work together very closely to produce the brief and then to create the building that solves the problem posed by the brief. Some clients have very clear ideas about their requirements, but it is the architect who interprets these ideas in the form of a design. If a client was able to sustain a claim to copyright in a design, it would have to be shown that the client took part in the transforming of the ideas into drawings, whether via drawing board or CAD machine or gave precise instructions as to what was to be included in the design.[107] In the rare case of a client being able accurately to draw a design that satisfied the brief and to pass it to the architect so that all that needed to be done was to draw it out neatly or to provide such clear instructions that the architect became, in effect, a draftsman, it might be that the client had a share in the copyright with the architect. However, that will be a very rare circumstance.

Even if the client gives the architect a detailed drawing of what is required, the architect will usually have to change it considerably in order to make it work in practice. It is difficult to show that one design has been copied from another, which is why there are relatively so few successful cases about infringement of copyright. Usually there have to be some significant features on both designs.

In the majority of instances, the architect retains copyright in the designs and the client has a licence to reproduce the design in the form of a building. Sometimes the terms of appointment expressly set this out as in the RIBA-produced forms of appointment. Even if the appointment document does not mention copyright, it will be implied that the client has a licence to reproduce the design if a substantial fee has been paid. As a rule of thumb, it is usually assumed that the licence will be implied if the client has paid for all work up to the end of RIBA stage D. If the fee paid is only nominal, no such licence will be implied. For example, a client will not normally have a licence to reproduce a design in the form of a building if the architect has been paid only for preparing a planning application.

107 *Cala Homes (South) Ltd v Alfred McAlpine Homes East Ltd* [1995] EWHC 7 (Ch).

79 If a designer has been paid for producing full drawings for a development and the client sells the site, can the new owner use the drawings to build on the site?

This is a question of copyright and, particularly, a question of copyright licence. Copyright legislation is enshrined in the Copyright, Designs and Patents Act 1988 as amended. Copyright matters are apt to be rather complex and proving copyright infringement is not straightforward. There is little doubt that if a designer has been paid in full for producing a complete set of production drawings, the client will have an implied licence to use such drawing to construct the building on that site. The terms of engagement used may vary that position of course, but certainly the provisions of the SFA/99, CE/99 and SW/99 and the RIBA Agreements 2010 expressly provide that the client has a licence to reproduce the design in the form of a building when full fees or a licence fee have been paid. The more complex question is the one posed here, when the site has been sold to a new owner.

In *Blair v Osborne & Tomkins (a firm) and Another*,[108] an architect designed a pair of houses and the site owners subsequently sold the site to others. The buyers commissioned other architects to develop the design and it was built on the site. An action for breach of copyright failed. The original architect was claiming that, when his appointment was terminated, even though he was paid in full to that stage, the client did not have a licence to use the drawings to build the houses. The Court of Appeal appear to have been greatly influenced by the existence of a clause allowing either party to terminate on reasonable notice. They considered that, if the architect was correct, it allowed him to hold the client to ransom. Perhaps more to the point, they also held that the licence which was implied in that case extended to avail the purchaser of the site. It is not entirely clear from the judgment why they did this. However, they were deciding the extent of the licence. They implied a licence for all purposes connected with the erection of the houses. It should be noted that Salmon LJ questioned the extent of the licence implied by this court in a subsequent case (*Stovin-Bradford v Volpoint Properties Ltd and Another*[109]).

108 (1970) 10 BLR 96.
109 (1971) 10 BLR 105.

If a licence is expressly stated, as in RIBA S-10-A for example, there should be no doubt about what it covers. In S-10-A, the client is said to have a licence to copy and use the material. Therefore, the licence applies to the client and to the client only. The general rule is that a licence cannot be assigned without permission, neither can rights under a licence be transferred without permission. Therefore, the client cannot transfer this licence to a third party without the designer's permission. But where terms of engagement simply say that the licence will be granted to construct the design on the site it appears that this licence can be interpreted to cover any person who is the successor in title to the property. The Court of Appeal in the *Blair* case appear to have taken the view that where nothing is expressly stated, the licence is to be given a wide interpretation. However, if the terms of engagement state that neither party will assign or transfer any of the benefits or obligations of the appointment without written consent from the other – a common provision – the situation is not so simple. The copyright licence is obviously a benefit of the appointment. Therefore, it is very arguable that the licence, although potentially applicable to all successors in title to the property, is not so applicable unless and until the designer consents to it in writing.

In summary, it may be said that where terms of engagement stipulate that the licence extends to the client only, a future purchaser of the site will not have a licence to use the design. If the terms are silent about the recipient of the licence, it is likely that the courts will imply a term that future purchasers of the site also have the benefit of the licence. However, future purchasers may be prevented from obtaining the licence if there is a clause which prohibits transfer of the licence without the designer's permission.

80 What does it mean to take 'reasonable skill and care' and how is that different from an obligation to provide something that is 'fit for purpose'?

The law requires a professional person to exercise reasonable skill in care in the performance of his or her duties. This standard is also required of the designer by the Supply of Goods and Services Act 1982 in respect of any contract for the supply of design services. This statutory duty can be displaced by the imposition of a stricter duty in a contract. The stricter duty is 'fitness for purpose'. The duty may be stated in a contract or the law will normally imply it where the contract is on the basis of work and materials unless it is clear that the employer is not relying on the contractor for design.[110] In *Greaves (Contractors) Ltd v Baynham Meikle & Partners*[111] the court said:

> The law does not usually imply a warranty that [an engineer] will achieve the desired result but only a term that he will use reasonable skill and care. The surgeon does not warrant that he will cure the patient. Nor does the solicitor warrant that he will win the case. But, when a dentist agrees to make a set of false teeth for a patient, there is an implied warranty that they will fit his gums, see *Samuels v Davis*.[112]

It is likely that differing standards will apply to different professions, and qualifications of a surgeon are not the same as those of an surveyor. Therefore, it may not be appropriate in every case to apply the common standard of the average competent professional. If a particular professional is very experienced or qualified it may be that the appropriate standard of skill and care will be the standard usually exercised by fellow members of the same profession having similar experience or qualifications.[113] The standard of care is judged at the time of the decision or act and not with hindsight, for example at the time of a hearing some years later when knowledge in the profession may have improved or changed. An important difference between reasonable skill and care and fitness for purpose is just that: the professional can use the standard of knowledge at the time of the act as a defence, whereas a contractor liable for constructing a building fit for purpose has little or no defence if the building does not perform properly and it is irrelevant that it can bring proof to show that it was not negligent. The obligation to use reasonable skill and care is an obligation related to the way the professional exercises the particular profession. The obligation to achieve fitness for purpose is not concerned with how something is done, but rather with the standard achieved when the 'doing' is finished.

110 *Young & Marten Ltd v McManus Childs Ltd* [1968] 2 All ER 1169.
111 [1975] 3 All ER 99.
112 [1943] 2 All ER 3.
113 *Duchess of Argyll v Beuselinck* [1972] 2 Lloyd's Rep 172.

81 Why does a design and build contractor usually have a fitness for purpose obligation, but not under the JCT Design and Build Contract?

Where an employer relies solely on a contractor to design and construct an entire building, a term of reasonable fitness for purpose will be implied and the contractor's liability will not depend on its negligence or fault or whether the unfitness results from the quality of work or materials or from defects in the design.[114] It has been expressed like this in a case concerning the collapse of a television mast:

> . . . in the absence of a clear, contractual indication to the contrary, I see no reason why [a contractor] who in the course of his business contracts to design, supply and erect a television mast is not under an obligation to ensure that it is reasonably fit for the purpose for which he knows it is intended to be used. The Court of Appeal held that this was the contractual obligation in this case and I agree with them. The critical question of fact is whether he for whom the mast was designed relied upon the skill of the supplier . . . to design and supply a mast fit for the known purpose for which it was required.[115]

Therefore, where a contract is entered into by which a contractor is required to design and construct a building, its liability will be to ensure that the building, when finished, is reasonably fit for its intended purpose so far as that purpose has been made known.

The JCT design and build contracts have amended this usual implication of law by inserting a special clause (DB clause 2.17.1) which expressly states that the contractor has the same liability as an architect or other appropriate professional designer. This is a valuable concession to contractors which is not available under most other design and build contracts.

114 Viking Grain Storage Ltd v T H White Installations Ltd and Another (1985) 3 Con LR 52.

115 Independent Broadcasting Authority v EMI Electronics Ltd and BICC Construction Ltd (1980) 14 BLR 1.

Chapter 11

Architect's instructions

82 What counts as an instruction?

This is a common question. Standard building contracts refer to instructions and whether they must be in writing or oral, how they may be confirmed and by whom, but contracts do not specify what constitutes an instruction. Usually, to qualify as a written instruction, there must be an unmistakable intention to order something and there must be written evidence to that effect. Not all written instructions are clear – some are decidedly vague (contractors might believe deliberately so). Although an instruction may be implied from what is written down, it is safer from the contractor's point of view to ensure that the words clearly instruct. To take a common example: a drawing sent to a contractor with a compliments slip is not necessarily an instruction to carry out the work shown thereon. It may be simply an invitation to the contractor to carry out the work at no cost to the employer, it may be inviting the contractor's comments or it may simply be saying: 'This is what we thought about doing, but we changed our minds.' Although most adjudicators would no doubt assume that a drawing sent with nothing but a compliments slip was an instruction to do the work shown on the drawing, such an assumption would be subject to challenge. All drawings should be issued with a letter or instruction form clearly instructing the contractor to construct what is on the drawing.

The same comment applies to copy letters sent under cover of a compliments slip. Architects sometimes send a letter to the employer saying that they are going to instruct the contractor to do certain additional work in accordance with the employer's wishes. Those same architects misguidedly believe that if they send a copy of that letter to the contractor, it amounts to an instruction to the contractor to get on with the work. Clearly, that is wrong. An instruction on a printed 'Architect's Instruction' form is valid if signed by the architect. An ordinary letter can also be a valid instruction. If the architect wishes, he or she can write the instruction on a piece of old roof tile or on the side of a brick. Providing they are signed and dated and legible, they are all valid instructions. The minutes of a site meeting may be a valid instruction if the contents are expressed clearly and unequivocally and particularly if the architect is responsible for the production of the minutes. However, site meeting minutes are obviously not a good medium for issuing instructions, because of the possible delay in distribution.

83 What can be done if a contractor refuses to carry out an instruction and refuses to allow the employer to send another contractor on to the site?

Clause 3.10 of SBC requires the contractor to comply forthwith (as soon as it reasonably can do so) with architect's instructions that are properly empowered by the contract. If the contractor refuses to do so or simply ignores requests to get on with the instruction, the architect is entitled to issue a written compliance notice under clause 3.11. This notice gives the contractor 7 days from receipt to comply with the instruction. If the contractor still refuses, the employer may employ others to do the work and then an appropriate deduction of all the additional costs may be made from the Contract Sum. So far so good. The question refers to the hopefully rare instance where a contractor refuses to give access to the site to the other contractor engaged by the employer to carry out the instruction.

In *Bath & North East Somerset District Council v Mowlem*,[116] the Court of Appeal was faced with an interesting conundrum. Many of the details are unimportant for this purpose, suffice to say that an impasse arose between the employer and the contractor because the contractor, having objected to and refused to carry out an instruction, would not allow the employer to bring another contractor on to the site to do it. The employer sought an injunction to prevent the contractor from refusing access. That is the background. It should be said that both parties must have believed that they had good reason for acting as they did up to that point.

116 (2004) 100 Con LR 1.

Now the position becomes interesting. Courts are generally reluctant to grant injunctions unless there is a true emergency. They will grant an injunction only if the problem is such that no amount of future damages can sufficiently recompense the injured party after trial. For example, a court may well grant an injunction to prevent someone chopping down a five-hundred-year-old oak tree in a prominent position because, once chopped down, no amount of money could restore the tree. However, a court would be unlikely to grant an injunction to prevent the demolition of an ordinary modern brick wall, because an award of money will certainly be enough to pay for its rebuilding.

The contractor argued that an injunction should not be granted because the contract contained a liquidated damages clause and, if the contractor was ultimately found to be wrong, the liquidated damages would recompense the employer for the resultant delay. The Court of Appeal disliked this argument. In granting the injunction to the employer, they decided that the contractor was in breach of contract for refusing access in this instance and that liquidated damages was not an agreement between the parties that the contractor could continue its breach of contract. Although liquidated damages was ordinarily the most damages that could be recovered for delay in completion, they did not properly compensate the employer for the loss it would suffer by the continuing breach.

On the basis of this case, it seems that employers can expect to obtain injunctions if contractors refuse access to the site to other contractors who have been lawfully engaged under the terms of the contract.

It is sometimes said that liquidated damages are not only damages due to the employer in the case of a breach on the part of the contractor to complete in time but are also to be regarded as the price payable by the contractor for the option of taking longer to complete. This case shows that such a view is not correct.

84 Can a contractor refuse to comply with an architect's instruction which requires the acceptance of the quotation of a sub-contractor chosen by the architect?

Architects like to be able to choose the identity of sub-contractors to carry out certain types of work, particularly where the sub-contractor is involved in the design. Lift installations, heating installations and piled foundations are examples of the kind of work. One of the reasons is that architects like to deal directly with these sub-contractors before the work is let to the main contractor in order to be able to integrate the sub-contract work with the Works as a whole. JCT 98 had complex provision for the architect to nominate sub-contractors, but nomination has been removed from the JCT successor contract SBC and only IC and ICD have provision for the architect to name sub-contractors although SBC retains provision for a short list of sub-contractors to be inserted in the bills of quantities from which the contractor is entitled to choose. Since either the employer or the contractor has the right to add further names to the list, it is often neither short nor particularly exclusive. Despite this, and whichever contract architects are using, they will very often attempt to tell the contractor the sub-contractors to be used. Architects do this in one of two ways: naming a firm in the bill of quantities or specification, or naming a firm in an architect's instruction, usually instructing the contractor to accept a quotation.

Naming a firm in the tender documents, which subsequently become the contract documents, will bind the contractor to use that firm unless, when tendering, the contractor qualifies the tender to exclude such firm. In the absence of such exclusion, the contractor has tendered to carry out the whole of the Works as described in the documents and, on acceptance, a legally binding contract comes into existence to that effect. The firm becomes a domestic sub-contractor for which the contractor has full responsibility in the normal way. If the documents purport to give the firm any design responsibility, it is essential that the main contract includes provision for such design responsibility or that there is a suitable warranty in place between the sub-contractor and the employer. A difficulty arises if the firm ceases work for some reason before the sub-contract work is complete. Such cessation may result from repudiation on the part of either contractor or sub-contractor or the firm may become insolvent or may be a result of termination procedures properly carried out under the terms of the sub-contract. In such circumstances, the contractor has neither the right nor the duty to engage another firm to complete the work. Strictly speaking, if the outstanding sub-contract work is other than trivial, the contract is frustrated and comes to an end. In practice, the contractor will usually call upon the architect to name a new firm to carry on the sub-contract work and the inevitable additional cost will be added to the Contract Sum.

The situation is much more difficult if the architect has not named a firm in the contract documents, but simply instructs the contractor to accept a quotation submitted at the invitation of the architect in respect of a provisional sum. Most standard form contracts give the architect power to instruct the expenditure of a provisional sum, indeed most contracts place an obligation on the architect to do so. The question is whether the contractor is obliged to comply with the instruction and accept the firm's quotation as instructed. There seems little doubt that where a contractor simply accepts the quotation, the firm whose quotation is so accepted becomes a domestic sub-contractor. However, if the architect has not ensured that the quotation includes everything necessary, any additional cost is to be added to the Contract Sum. Some contractors who accept such quotations attempt to argue that the sub-contractor is thereby a nominated sub-contractor with all the consequences which would result under JCT 98. Although there is one recorded case where a court was prepared to treat a sub-contractor as nominated even though not nominated properly under the contract provisions, that was only because both the parties concerned had been treating the sub-contractor in that way.[117] It is thought that if the firm has not been named or listed in the contract documents, the contractor can refuse to accept the quotation. It may be that the contractor would have to show that the refusal was reasonable, but it should not be difficult for a contractor to put forward a reasonable ground for refusal: such things as the contractor's past bad experience of that particular firm or the fact that it had heard bad things about it. Indeed it is probably not going too far to say that a valid reason for refusal would be for the contractor simply to say that it was not prepared to take on, as a domestic sub-contractor, a firm which it had not chosen and for which it was not prepared to take responsibility.

85 Should AIs be signed by an individual or the firm?

This question crops up from time to time. It is usually asked by architects fearful that an instruction will be invalid if not signed by the correct person.

117 *St Modwen Developments Ltd v Bowmer & Kirkland Ltd* (1996) 14-CLD-02-04.

The simple answer to the question is that an AI may be signed by any person who is authorised to do so. The architect is the person named in the contract. Only the architect may issue certificates and instructions under the terms of the contract, but that necessarily includes anyone authorised by the architect. The architect should be careful to inform all interested parties of the names of all persons authorised to act on behalf of the architect.

Very often, the name of the architect in the contract will be a firm 'XYZ Architects' or some such name. Therefore, the letter informing all parties of authorised persons, must be signed by 'XYZ Architects'. If the firm is a limited company, the signature of a director will do: if a partnership, it should be one of the partners. If it is a limited liability partnership, it ought to be one of the designated members.

Where AIs, certificates or letters are signed by an authorised person, that person should sign 'for and on behalf of'. This is undoubtedly the best method. It is not sufficient that the letter, etc. is on headed paper. The important thing is that it must be plain that the signatory is not signing on his or her own behalf, but on behalf of the architect, be that company, partnership or sole principal. So it is probably sufficient if the name of the firm is typed where the signature would normally go and the authorised person signs immediately underneath. Sometimes people sign the actual name of the architect. For example, if the named architect is John Smith, one of the authorised persons, say Alice Davis, may sign 'John Smith' provided she initials the signature.[118]

The use of mechanical impressions of signatures are of doubtful validity and should be avoided. A mechanically impressed signature together with the initials or signature of an authorised person is probably valid.

86 If the employer gives instructions on site directly to the contractor, must the architect then confirm those instructions in writing?

Many employers seem to find it difficult to stay away from site. It should go without saying that employers should never be allowed to visit site unaccompanied. At best they will get a warped idea of what

118 *London County Council v Vitamins Ltd* [1955] 2 All ER 229.

is happening (for example, all the rooms look too small at foundation stage); at worst they may answer questions from the contractor or give instructions even when no questions are asked. The contractor should be carefully briefed at the pre-start meeting always to refer any queries to the architect and never to ask the employer any questions directly. Despite this, instructions may be given directly to the contractor and the contractor may carry out the instruction without reference to the architect. The architect is not obliged to confirm the instructions in writing.

The first thing to establish is why and in what circumstances the instructions were given. The second is to establish the effect of the instructions on the Works as a whole. It may be that the instruction was given by the employer who told the contractor to check with the architect, or it may be that the employer gave the instruction without really understanding what was being asked. Neither of these circumstances exonerates both employer and contractor from the charge of failing to act in accordance with the contract of course, but life is like that. People fail to act as they should.

If the architect decides that the instruction, although given directly, is simply the kind of instruction that, if the employer had asked the architect to issue, would have been issued without difficulty, the architect will presumably have no problems with ratifying the instruction. The position becomes more difficult if it is an instruction that the architect would not have issued and which perhaps has a detrimental effect on the project. There is no doubt that the employer and contractor, as parties to the contract, are entitled to vary the terms as they wish. If the employer decides to give a direct instruction, albeit the contract provides for only the architect to do that, and if the contractor accepts the instruction, it is likely that either a fresh little contract has been formed for that item of work or, alternatively, it may rank as a variation to the original contract. Obviously, the architect cannot include the value of such a variation in a certificate unless it is the subject of an architect's instruction. If the architect does not confirm with an instruction, the cost of the variation must be paid by the employer directly.

A contractor who accepts a direct instruction from the employer is unwise. If the contractor carries out the work, but the employer contends that the instruction was never given, the contractor is in breach of contract and can be obliged to amend the work to conform to the contract documents.

87 Does the architect have power to give instructions after practical completion?

There appears to be no reason why the architect should not be empowered to issue instructions after practical completion subject to certain exceptions. SBC clause 3.10 permits the architect to issue instructions in regard to any matter about which the contract expressly authorises instructions. Unlike the position under other contracts such as ACA 3 and GC/Works/1, the JCT contracts do not list permissible instructions conveniently in one clause. One has to search through the contracts. For example, in SBC, there are twenty-one clauses throughout the contract which authorise instructions. Once such clauses have been identified, the question is whether the architect's power to issue them expires at practical completion.

None of the clauses is specific on this point, but it is obvious that instructions under some clauses cannot be issued once the Works have been completed. It is plain that instructions requiring variations to the Works cannot be issued after practical completion, because practical completion is when the Works are complete save for minor things still to be done. The issue of an instruction requiring substantial changes would be inconsistent with that. Moreover, practical completion marks the beginning of the period during which the contractor is to provide information for the preparation of the adjusted Contract Sum. If there was no stop to variations, the contractor would not know when the final account documentation could be completed. There is judicial opinion in support of this view.[119] A subsequent decision to the contrary is sometimes quoted.[120] However, in that case a JCT prime cost contract was under consideration. There was no stipulated scope of work and the contractor simply carried out instructions. Moreover, the contractor in that case agreed to remain on site to execute the instructed work. On the same basis, the architect will have no power to postpone work after practical completion, no power to issue instructions about setting out, drawings or discrepancies. However, there seems no reason why the architect should not be able to issue instructions regarding defects, making good and requiring opening up or testing.

119 *New Islington and Hackney Housing Association Ltd v Pollard Thomas and Edwards Ltd* [2001] BLR 74.

120 *Treasure & Son Ltd v Martin Dawes* [2007] EWHC 2420 (TCC).

88 What is the position under a JCT traditional contract if the contractor acts on instructions given directly by the mechanical services consultant?

The JCT traditional contracts are SBC, IC, ICD, MW and MWD. The parties to these contracts are the employer and the contractor. There is provision for naming an architect or other contract administrator. SBC clause 3.10, IC and ICD clause 3.8, and MW and MWD clause 3.4 require the contractor to comply forthwith with any instructions, authorised under the contract and issued by the architect. There is no provision in any of these contracts for the mechanical services consultant or anyone else to issue instructions. Therefore, an instruction with which the contract must comply cannot be issued by anyone other than the contract administrator. Obviously, as parties to the contract, the employer and the contractor may agree to vary some term or some piece of work; but the contractor is entitled to refuse to carry out any instructions given by the employer.

The mechanical services engineer may be a sub-consultant to the architect or, more usually, engaged directly by the employer. In either case, neither the engineer nor any other consultant is entitled to give instructions to the contractor under the terms of any of these contracts. The correct procedure is for such consultants to give the architect a note of the instruction required and, when the architect is satisfied about the need for such instruction it is a matter for the architect to put it into an architect's instruction issued under the terms of the contract. Therefore, a consultant who gives a direct instruction to the contractor is attempting to operate in breach of the terms of the building contract. A contractor which accepts such instructions and proceeds to carry them out is in breach of its obligations to construct the Works in accordance with the contract documents. The architect is entitled to instruct the contractor to remove such work as being work which is not in accordance with the contract. If the work, although not properly instructed, is nevertheless work which should have been carried out, the architect has power under the contract to ratify such instructions or, if appropriate, to accept the work with a consequent reduction in the Contract Sum.

Sometimes, the architect, at an early site meeting or in some other way, will say that consultants may visit site independently and may issue instructions directly to the contractor. This is a thoroughly bad practice and sets the scene for serious misunderstandings and other problems. The effect is probably to make each of the consultants the authorised representative of the architect for that project. Architects who make this kind of arrangement have a death wish.

89 Does the wording of MW give the architect power to issue an instruction to postpone the Works?

The main provision governing the issue of instructions under MW is clause 3.4. The clause is broadly worded, giving the architect power to issue instructions with which the contractor must comply forthwith. At first sight, it appears to give the architect power to issue instructions about anything connected with the contract. However, the courts generally look to restrict such broad wordings and it is likely that it will be interpreted so as to allow the architect to give such instructions as are necessary for the proper administration of the contract. There are only five other clauses which refer to instructions:

- Clause 2.10 permits the architect to issue instructions that the contractor is not to make good defects.
- Clause 3.6 permits the architect to order variations.
- Clause 3.7 requires the architect to issue instructions about the expenditure of provisional sums.
- Clause 3.8 permits the architect to issue instructions for the exclusion from site of any person employed on the Works.
- Clause 5.4B.2 requires the architect to issue instructions for reinstatement after insured damage.

It is clear that there is no express power to issue instructions postponing any part of the work. However, in the case of a relatively short contract such as MW, the courts may well imply terms if they are necessary to make the contract effective or they will construe the existing terms of the contract, such as clause 3.4, in a purposive way. For example, although not expressly stated, it is considered that the architect must have power under clause 3.4 to issue instructions requiring the removal of defective work and materials. Allied to that power is the power to order opening up and testing. It is thought that architect probably has that power and that it would be an implied term that where the opening up or the test results showed that the work was in accordance with the contract, the employer would be obliged to reimburse the contractor for its costs and, if appropriate, an extension of time must be given, because the delay was beyond the contractor's control. The architect must have power to issue instructions for the correction of inconsistencies, otherwise the first discovered inconsistency could bring the project to a halt.

It is also likely that the architect would have power to postpone the whole or any part of the Works. The power would not extend to differing possession of the site. It seems inevitable that any postponement would result in an extension time as being outside the contractor's control. However, additional payment to reimburse the contractor for associated costs may not be simple. It could not be claimed under the contract provisions, because there is no provision for ascertainment of loss and/or expense in those circumstances. Moreover, it could not be claimed as damages for breach of contract because, if the architect does have power under the contract to postpone work, the exercise of such power cannot be a breach of contract. It appears that the only way that the contractor could be reimbursed would be by the implication of another term to that effect. That in fact seems the sensible and just solution.

90 Can the clerk of works stop the Works?

Clerks of works are sometimes heard to say that they have 'put a stop order' on the Works or that they have 'stopped the job'. More commonly, they are heard to threaten these things. A 'stop order' sounds suitably impressive and quite official. For contractors and architects, who are not as familiar with building contracts as perhaps they should be, the threat can sound intimidating and reassuring respectively; the contractor threatened and the architect reassured that the clerk of works knows exactly what to do and is not afraid to do it. The reason for the threat is not important, but it may be because the contractor is not progressing to the satisfaction of the clerk of works in rectifying defective work. The powers and duties of the clerk of works are set out in SBC clause 3.4, and IC and ICD clause 3.3. All these clauses state that the clerk of works is acting solely as inspector on behalf of the employer and under the direction of the architect. SBC gives the clerk of works the additional power to issue directions provided they are about matters which the architect is expressly empowered to issue instructions and provided that the architect confirms such directions in writing within 2 days – which rather blunts the power.

Any search through the provisions to find reference to the clerk's of works power to stop the Works will be in vain for the simple reason that a clerk of works does not have such power under any of the standard form building contracts. Indeed, the idea that the clerk of works can issue an ultimatum and then stop the Works is ludicrous. The architect may postpone the whole or any part of the Works and, after suitable preliminary notice, the employer may terminate the contractor's employment if the right criteria are satisfied, but the clerk of works may only inspect and report to the architect. Contract administration is in the hands of one person: anything else would lead to chaos. A clerk of works who threatens to stop the Works clearly has little idea of what to do and if the clerk of works attempted to stop the Works, the action would be seriously in excess of his or her powers. A contractor which stopped the Works on being instructed to do so by the clerk of works would be in breach of contract and, if the stoppage was prolonged, in danger of repudiating its obligations under the contract.

Chapter 12

Inspection

91 What is the architect's site inspection duty?

Inspection is not something to be carried out lightly. Many architects simply wander on to the site with no very clear idea of what they expect to find, nor indeed what they should be looking for. The RIBA Agreements do not deal in detail with inspection of the Works. They simply require the architect to make visits to the construction Works in connection with general inspection of progress and quality of work, for the approval of any elements reserved for the architect's approval, obtaining information necessary for the issue of notices, certificates and instructions at intervals reasonably expected to be necessary at the date of the appointment and to advise if a clerk of works is necessary. It must be recognised that the number of visits is only an estimate.

It is perhaps cynical to say that a court will find that an architect's duty is to find just those problems that have been missed.

Nevertheless, architects may have difficulty in ensuring that they are not open to legal action from their clients for failure to inspect adequately. The best safeguard for any architect is to be able to demonstrate to a court that their inspection duties were carried out in an organised manner, having regard to what the courts have said. Therefore, before commencing an inspection of the Works, the architect must have a plan of campaign as follows:

- Inspections should have a definite purpose. They should coincide with particular stages in the Works. It is sensible for the architect to sit down beforehand and draw up a list of parts of the construction that must be inspected on that particular visit, together with items of secondary importance to be inspected if possible. The composition of the list and the frequency of inspections will depend on factors such as the employment of a clerk of works, the size and the complexity of the project and the experience and reliability of the contractor. Comments can be made against the checklist as the inspection progresses. The list and the comments are for the architect's own files, not for distribution. Although an architect's inspection duties are quite onerous, he or she will be better able to defend themselves in court against an allegation of negligent inspection if they can show, by reference to contemporary notes, that inspections were carried out in an organised manner.[121]

- Times of inspections should be varied so that a devious contractor cannot rely upon concealing poor work between inspections.
- The architect should always finish an inspection by spending a few minutes inspecting at random.
- Action should be taken immediately the architect returns to the office, whether or not any defects have already been pointed out to the site manager. It is wise to put in writing all comments regarding defective work.
- During site inspections, the architect is bound to be asked to answer queries. It is prudent to give answers on return to the office when it is possible to sit down and calmly assess the situation. Many decisions made on site are either amended or regretted later.

An architect's failure to inspect will not excuse a contractor from maintaining proper quality control systems. The contractor has undertaken to carry out the Works in accordance with the contract, not carry out the Works to as low a standard as possible unless the architect notices. Nevertheless, although the architect owes no duty to the contractor to find defects,[122] the client is entitled to expect the architect to carry out such inspections as will identify serious defects and a reasonable proportion of minor defects.

If a detail is more complex than usual, the architect will be expected to take more care in inspecting. Just because it is difficult to

121 *East Ham Corporation v Bernard Sunley & Sons Ltd* [1965] 3 All ER 619; *Sutcliffe v Chippendale and Edmondson* (1971) 18 BLR 149; *Brown & Brown v Gilbert Scott & Payne* (1992) 35 Con LR 120; *Alexander Corfield v David Grant* (1992) 59 BLR 102; *Bowmer & Kirkland v Wilson Bowden Properties Ltd* (1996) 80 BLR 131.

122 *Oldschool v Gleeson Construction Ltd* (1976) 4 BLR 103.

inspect something does not mean that inspection is not necessary. It is even more necessary, because the contractor might be relying on the difficulty of inspection to attempt to get away with defective work. If work, by its very nature, is being covered up almost as soon as it is done, the architect might argue that there is little point in inspecting because, although the operatives will carry out the work properly while the architect is there, as soon as the architect goes, they will revert to poor workmanship. In fact, that is a cogent reason for continuous inspection of that particular element.[123] The architect's knowledge of the skill and experience of the contractor is an important factor; more time must be spent inspecting the work of an inexperienced contractor. It will usually avail the architect nothing to say that reliance was placed on the contractor's assurance that everything had been properly executed. But if an architect has a great deal of experience of the work of a particular contractor and knows it to be good, reliable and conscientious, less inspection should be needed.

The number of visits and their duration is not the test of adequate inspection. The key is the number of visits necessary. Therefore, it is no defence for an architect to say: 'I visited twice a week for an hour each time.' It may be that in some instances that is unnecessary: in others it is too few or too short. Moreover, an architect should expect to have to visit site more often at some stages of the work, occasionally spending full days if very important work is being done. The leading authority on the architect's duty to inspect is the decision of the House of Lords in *East Ham Corporation v Bernard Sunley & Sons Ltd*:

> As is well known, the architect is not permanently on the site but appears at intervals, it may be of a week or a fortnight, and he has, of course, to inspect the progress of the work. When he arrives on the site there may be very many important matters with which he has to deal: the work may be getting behind hand through labour troubles; some of the suppliers of materials or the sub-contractors may be lagging; there may be physical trouble on the site itself, such as, for example, finding an unexpected amount of underground water. All these are matters which may

123 *George Fischer Holdings Ltd v Multi Design Consultants Ltd and Davis Langdon & Everest* (1998) 61 Con LR 85.

> call for important decisions by the architect. He may in such circumstances think that he knows the builder sufficiently well and can rely upon him to carry out a good job; that it is more important that he should deal with urgent matters on the site than that he should make a minute inspection on the site to see that the builder is complying with the specification laid down by him ... It by no means follows that, in failing to discover a defect which a reasonable examination should have disclosed, in fact the architect was necessarily thereby in breach of his duty to the building owner so as to be liable in an action for negligence. It may well be that the omission of the architect to find the defects was due to no more than an error of judgment, or was deliberately calculated risk which, in all the circumstances of the case, was reasonable and proper.

These are comforting words, but it is important to give them due weight in the light of the other decisions.

92 What is the position if the contractor has covered up work?

There are two possible reasons why there may be a problem with work covered up. The first is that the work may be defective and, if covered up, the defects may not make themselves known until after the employer has taken possession of the completed building with all the consequent disruption while remedial work is undertaken. The second reason is that some of the work covered up may need measuring in order to value and certify.

The problem of defective work being covered can be dealt with in two ways. The best way is to have a clause inserted in the contract which makes clear that no work may be covered until the architect or the clerk of works has inspected it and that if covered before it is inspected, the contractor must, without charge to the employer, uncover the work to the extent required by the architect in order to carry out the necessary inspection. Preambles to bills of quantities commonly contain a clause to that effect. The downside to this is that the smart contractor will probably require the architect to express satisfaction in written form with work about to be covered on the perfectly reasonable premise that it needs the written statement to avoid the possibility that the architect might subsequently deny having had the opportunity to inspect. Many contractors will give the architect 2 or 3 days, notice that something is to be covered. If reasonable notice is given, it will be difficult to blame the contractor if the architect does not inspect. Alternatively, the contractor can interpret the clause strictly and refuse to cover the work until the architect has inspected even if there is a resultant delay to the progress of the Works. A good compromise is to have a clause written into the contract which requires the contractor to give the architect a reasonable period of notice, say 24 or 48 hours, before covering any work. Whether or not the architect inspects, inspection may be deemed to have occurred.

The other way to deal with the problem is for the architect to issue an instruction requiring the contractor to open up for inspection any work which has been covered up. Most standard form contracts allow for that kind of instruction (for example SBC clause 3.17). Unfortunately, if the uncovered work is in accordance with the contract, the employer has to pay for the opening up and the contractor may be entitled to loss and/or expense and an extension of time. This method requires the architect to be fairly sure there is something wrong before issuing an instruction.

It is usually in the contractor's interests to have work measured which is about to be covered up, particularly if the work carried out is additional to what is described in the bills of quantities. Therefore, where a contractor covers up work which is extra to what is measured without notifying the quantity surveyor first, the contractor will only have itself to blame if the quantity surveyor declines to value the supposed extra work. Indeed, it is difficult to see how the quantity surveyor can do other.

Therefore, if the contractor has covered up work and there is no additional clause in the contract dealing with the position, the architect will be obliged to make a decision about opening up the work with the possibility of additional costs if it is found to be in accordance with the contract. However, if a clause is included and if the contractor covers up work either before the architect has inspected or before the notice period has run out or without giving notice at all, depending on the wording of the clause, the contractor will be in breach of contract. The employer will be entitled to damages for the breach – sufficient to put the employer in the same position as if the breach had not occurred. In other words, the employer could engage others to open up and check the work before covering over again. In practice, the contractor would probably realise its error and open up and reinstate itself; obviously it would not be entitled to payment, whatever was found, because it would be remedying its own breach.

93 Is there a difference between inspecting and supervising?

'Inspection' and 'supervision' are often confused. Architects are commonly referred to as being responsible for 'design and supervision'. This is to confuse supervision with inspection. Supervision is obviously a more onerous obligation than inspection and one which can only be carried out by someone who has control over the workforce. In *Brown & Brown v Gilbert Scott & Payne* the court said:

> In my judgment the [architect] had a duty to inspect the works of the [contractor] and that the use of the word 'supervision' does not enlarge his duty in any way. As was said in evidence by one of the experts, 'supervision' was the word which used to appear in the RIBA Form of Engagement but in more recent editions this is replaced by the obligation to 'inspect'.[124]

That appears to be a wrong view. Inspection involves looking and noting, and possibly carrying out tests. Supervision, however, not only covers inspection, but also the issuing of detailed directions regarding the execution of the Works. Supervision can be carried out only by someone with the requisite authority to ensure that the work is undertaken in a particular way. That is the prerogative of the contractor. A subsequent case decided that 'inspection' is a lesser responsibility than 'supervision'.[125] That is the better view.

94 Is the architect liable for the clerk of works' mistakes?

The answer to that question depends, in part at least, on whether the clerk of works is employed by the architect or by the client.

In *Kensington & Chelsea & Westminster Area Health Authority v Wettern Composites*,[126] the employer had engaged the clerk of works

124 (1992) 35 Con LR 120.
125 *Consarc Design Ltd v Hutch Investments Ltd* (1999) 84 Con LR 36.
126 (1984) 1 Con LR 114.

and was responsible for his payment. The employer took proceedings against the architects, because some of the precast concrete mullions were found to be defective. The court held that the presence of a clerk of works did not remove or reduce the architects' obligation to use reasonable skill and care in inspecting the work, but the employer was vicariously liable for the negligence of the clerk of works although the clerk of works was under the architects' direction. The architects were found liable, but their damages were reduced by 20 per cent to take account of the employer's liability through the clerk of works.

The importance of the clerk of works being employed by the employer was noted in passing in *Gray (Special Trustees of the London Hospital) v TP Bennett & Son.*[127] This was a case where defects were discovered in the brickwork and supporting concrete nibs of a nurses' home some 25 years after it had been built. None of the professionals were found to have been negligent and the defects were found to have been deliberately concealed. In regard to the architect and the clerk of works, the court said:

> . . . it is clear that Mr Potts, as clerk of works, was to be the employee of the hospital, even though recommended by the architects. Furthermore, on the evidence it was established that at the end of the job, it was the Bursar of the hospital who notified him that his employment was at an end. I appreciate that the question of control frequently determines who in reality is utilising his services for the purpose of establishing vicarious liability, but in this instance the architect from the outset required an indemnity from the hospital if they were to accept him as an employee. That indemnity the hospital were not prepared to give, with the result . . . that the clerk of works remained their employee in the capacity of inspector for the building owner as laid down in . . . the contract, although he was also the eyes and ears of the architect.

As an employee of the hospital, the architect was clearly not liable for the clerk of works' actions, albeit in this instance the clerk of works was found to have carried out his duties strictly and no blame whatsoever was attached to him.

127 (1987) 43 BLR 63.

95 What is the position if the clerk of works approves defective work?

Under JCT contracts, the clerk of works has no power to approve anything, nor to give instructions to the contractor. IC and ICD, under clause 3.2, expressly restrict the power of the clerk of works to inspecting on behalf of the employer but under the direction of the architect. SBC, clause 3.4, allows the clerk of works to issue directions, but only in regard to things about which the architect is empowered to issue instructions. Moreover, such directions are stated to be of no effect unless confirmed in writing by the architect within 2 days. In practice, of course, many contracts would come to a complete stop if clerks of works stopped giving instructions to contractors and if contractors stopped complying. Nevertheless, the position remains that a contractor which complies with an instruction of the clerk of works does so at its peril.

Clerks of works are sometimes called 'the eyes and ears of the architect' and rightly so. A good clerk of works will be on site at regular intervals and, on larger projects, sometimes permanently. The clerk of works will usually strike up a relationship with the contractor which should be one of mutual respect rather than friendship. If the clerk of works is permanently on one site, it is very difficult to keep the kind of reserve necessary. Most architects will make every effort to support the clerk of works, but sometimes a clerk of works will go too far. The nightmare scenario for an architect to arrive on site to inspect the work and to instruct the site agent that certain work is unsatisfactory only to be informed that the clerk of works has already approved or accepted it. There is no easy way of dealing with that situation. The architect has no alternative but to make plain to the contractor the limits of the clerk of works' authority and to refer to the contract. An architect's instruction must be issued requiring the removal of the defective work and a quiet word with the clerk of works is indicated. The result is likely to be that the contractor will subsequently refuse to take any instructions or directions from the clerk of works even of a clarifying nature. That may result in the project suffering some delays.

In order to reduce the chance of this kind of situation developing, it is suggested that the powers and duties of the clerk of works are set out during the pre-start meeting and confirmed in the minutes. Of course, it is always open to the parties to agree that the clerk of works may give instructions and the contract can be amended accordingly. However, the consequence is likely to be confusing for the contractor and frustrating for the architect. In practice, there can only be one source of instructions and logically that is the architect.

Chapter 13

Defects during progress

96 Can the architect stipulate when the contractor must rectify defective work under SBC or can the contractor simply leave it all until just before practical completion?

During the progress of the Works, the architect is given powers to deal with defects in SBC clause 3.18. There are basically two kinds of defects: those due to an inadequate specification that are not the contractor's problem; and those due to work not being in accordance with the contract. It is only the second kind with which the contract is concerned. The architect may issue instructions regarding the removal from site of any defective work, goods or materials. Nothing in the clause entitles the architect to instruct when the defects must be corrected. This is in accordance with the contractor's right to plan and perform the Works in whatever way it chooses.[128]

If, in the opinion of the architect, the contractor does not comply within a reasonable time with an instruction to rectify work not in accordance with the contract, the architect has two possible ways to approach the difficulty. Clause 3.11 gives the architect power to issue a notice to the contractor, giving it 7 days from receipt in which to comply with an instruction. If the contractor fails to comply, the employer may engage another contractor to carry out the instruction and the original contractor will be liable for all the additional costs incurred by the employer, which must be deducted from the Contract Sum. Such additional costs will, of course, include any additional professional fees charged to the employer as a

128 *Greater London Council v Cleveland Bridge & Engineering Ltd* (1986) 8 Con LR 30.

result of the contractor's failure. This will be the route of choice in most cases – assuming that a couple of threatening letters do not do the trick first.

As a last resort, the architect may send a default notice to the contractor under clause 8.4.3 giving notice that the employer may terminate the contractor's employment if it refuses or neglects to comply with the architect's instruction to remove defective work and, as a result, the Works are materially affected. This ground used to be qualified by the word 'persistent'. That is no longer the case; the important point is that the Works must be substantially affected. The particular ground appears to be aimed at defects that are about to be covered up or which, for some other reason, would be awkward to put right if not given prompt attention. Therefore, if there is no urgency about the need to make good, this remedy is not appropriate and the contractor is entitled to plan the making good to fit in with its other work.

The position is, therefore, that in principle the contractor is entitled to plan its work, including making good, to suit itself. However, the architect is always entitled to insist on compliance with an instruction within 7 days. In serious cases where the integrity of the Works is threatened, termination can be considered.

97 The contractor incorrectly set out a school building, but it was not discovered until the end of the project when floor tiles in the corridor were being laid. What should be done?

Much depends on the effect of the incorrect setting out. If it resulted in the school encroaching over the boundary on to another person's land, it is virtually certain that, unless a deal can be done with the adjoining owner, the offending part of the school would have to be taken down and rebuilt to a different design. This could be very expensive for the contractor, if indeed the problem was incorrect setting out rather than incorrect setting-out drawings.

There are other possibilities. For example, the school might simply have gained half a metre in length, but it might cause no one a problem. In such circumstances, the school authorities have more school to heat and light, but against that, there is slightly more accommodation. If the gain is minor and of no consequence, it is technically a breach of contract, because that particular part of the Works is not in accordance with the contract, but both parties are likely to let the matter rest. Obviously, the client will not be prepared to pay for the extra walls, floors and roof and they should not be valued.

A trickier difficulty arises if the poor setting out results in the loss of half a metre or the awkward internal arrangement of part of the school. One question the client is sure to ask is why the error was not picked up sooner by the architect. If the error resulted in an internal planning problem it is indeed difficult to see why it was not picked up earlier than when the floor tiles were laid. When the error is picked up only at that late stage, it suggests that it is purely one of length or breadth and it is only when the floor tiling pattern is disturbed that it becomes apparent. It will be for the architect, if challenged, to provide evidence that site inspections were properly carried out and that the average architect in that position would not have found the error.

The basic contract position is that an employer is entitled to get what is being paid for. If I pay for blue boxes, that is what I should have and not green boxes even though the colour may not matter to anyone but me. Alongside that is the rule that where there is a breach of contract the injured party is entitled, so far as money can do it, to be put back in the position it would have had if the contract had been properly performed.[129] However, the courts have modified that rather tough position and they will take all factors into account before agreeing that a contractor in this position must spend large sums of money. The principle the courts apply is that the benefit provided by the remedial work must outweigh the cost of putting it right.[130] In some instances this is easy to calculate. In the *Ruxley* case, a swimming pool was not built deep enough at the shallow end, but the House of Lords decided that it was not worth the cost of demolishing and reconstructing the pool.

In that case, a factor was also whether the injured party would use the money to reconstruct. If something is without value or seriously reduced in value by the error, it is likely that a court would uphold its replacement. On the other hand, it is unlikely that a court would instruct wholesale demolition purely on aesthetic grounds. Therefore, although the baseline is that the contractor is responsible for its errors, care should be taken if the errors are expensive to correct for little apparent benefit.

129 *Robinson v Harman* (1848) 1 Ex 850.
130 *Ruxley Electronics and Construction Ltd v Forsyth* [1995] 3 All ER 268.

98 Is the contractor responsible for rectifying defects which the architect has noticed, but failed to report?

The contractor's obligation under standard form building contracts is to carry out and complete the Works in accordance with the contract documents. This obligation does not depend on whether the architect or the clerk of works spots errors and, in theory, the architect could go on holiday until the completion date provided all the necessary information for construction had been provided to the contractor. A defect is simply a breach of the contractor's duty to construct the Works in accordance with the contract documents. Therefore, it follows that the contractor should construct the building without any defects at all. If there are defects, the contractor has a clear duty to rectify them. Whether or not such defects have been seen by the architect is irrelevant. The architect has no duty to find defects for the contractor.[131]

That said, it is obviously bad practice, to say the least, for an architect to notice a defect, but to fail to point it out to the contractor. Even so, the contractor generally has no remedy because, as already stated, the architect owes no duty to the contractor to point out defects. However, care must be taken when using SBC, because clause 3.20 states that in the case of any materials, goods or workmanship comprised in work which, under clause 2.3, is a matter for the architect's reasonable satisfaction, the architect must give reasons for dissatisfaction within a reasonable time of the execution of such work. It appears to have been held that all work is inherently a matter for the architect's reasonable satisfaction.[132] The effect of clause 3.20 seems to be to amend the common law position and require the architect to inform the contractor of defective work. Clearly this clause is a likely candidate for deletion when contracts are being drawn up.

99 What can be done under SBC if a serious defect arises when the Works are nearly finished if the contractor denies liability and the employer is desperate to move in?

This is the kind of question which crops up frequently and to which there is no answer which will completely satisfy the employer. Here

131 *Oldschool v Gleeson Construction Ltd* (1976) 4 BLR 103

132 *Crown Estates Commissioners v John Mowlem & Co Ltd* (1994) 70 BLR 1 (CA).

is a situation where the Works are on the brink of being certified as achieving practical completion. Practical completion cannot be certified if there is a known defect present. Therefore, practical completion cannot be certified in this instance. It is established that the contractor has an implied licence to occupy the site of the Works until practical completion. Therefore, the employer is precluded from resuming occupation. To compound the problem, the contractor is denying liability for the defect and there is no immediate prospect of a resolution to the difficulty.

In order to see a way forward it is important to break down the several parts of the problem. Contractually, the key is the serious defect. That must be rectified before practical completion can be certified. The first and obvious step is for the architect to issue an instruction under clause 3.18.1. This somewhat awkwardly worded clause refers to the removal from site of work, materials or goods which are not in accordance with the contract. This is the clause the architect must use in order to secure the correction of the defect. Presumably the architect has already issued such an instruction, because the contractor has made clear that it does not intend to remedy the defect, at any rate unless it is paid. The architect should issue a notice under clause 3.11, which can simply be in the form of a letter mentioning the clause, requiring the contractor to comply with the earlier instruction within 7 days of receipt. Because receipt is the important occurrence, the notice is best sent by special delivery post which will ensure delivery by the next business day. If the contractor complies, all is well. If it does not comply, the employer can then engage another contractor to rectify the defect. The contractor will be liable for all the costs incurred by the employer and these costs can be deducted from the Contract Sum. The architect can then immediately issue the certificate of practical completion and the employer can move into the building.

On the assumption that the defect, although serious, would not pose a danger to health or safety, the employer could take matters in hand and ask the contractor to consent to partial possession of the whole of the Works. On receipt of such consent, the employer would simply retake possession and practical completion of the Works would be deemed to have occurred.[133] That would still leave the

133 *Skanska Construction (Regions) Ltd v Anglo-Amsterdam Corporation Ltd* (2002) 84 Con LR 100.

problem of the serious defect, but it could be dealt with following occupation by issuing an instruction followed by a compliance notice. A problem with this approach is that after practical completion of the Works (deemed or otherwise) the architect must issue an interim certificate releasing the first half of the retention (clause 4.20). If the release would leave the employer with insufficient funds to deal with the serious defect, it may be necessary for the employer to issue a withholding notice in respect of the interim certificate for cost of rectifying the defect. It should be noted that when Part 8 of the Local Democracy, Economic Development and Construction Act 2009 commences it will amend the Housing Grants, Construction and Regeneration Act 1996 requirements for notices before payment.

It is unlikely that the contractor would refuse consent, first, because the contract provides that such consent must not be unreasonably delayed or withheld, and second, because deemed practical completion starts the rectification period, it brings the contractor's Works insurance to an end and it ends the period during which liquidated damages can be incurred.

Chapter 14

Defects after practical completion

100 The contractor has re-laid a defective floor at the end of the rectification period. Can the cost of relaying the carpet be deducted from the final account?

Under SBC clause 2.38 and similar clauses under other JCT contracts, the contractor is entitled to return to site to make good those defects notified to it in the schedule of defects delivered to the contractor by the architect at the end of the rectification period. The defects in question are defects, shrinkages and other faults that are due to materials or workmanship not being in accordance with the contract or a failure by the contractor to carry out its obligations under the contractor's designed portion. An appropriate deduction is to be made from the Contract Sum in respect of those defects which the architect, with the employer's consent, has instructed the contractor not to make good.

It is assumed that the carpet to which the question refers has been purchased and laid by the employer after practical completion of the Works. The defects are breaches of contract on the part of the contractor. The question is whether the contractor is liable for the cost of having the carpet professionally re-laid after the remedial work. The answer to the question depends on the principle of foreseeability. In other words, at the time the contract was executed, was it obvious to the contractor that, if a defect arose in the flooring which had to be put right by the contractor completely relaying the floor, it was likely that the employer would have a carpet laid of the same type and quality as was in fact the case? If the answer to that question is 'Yes', the contractor is liable for the cost of relaying the carpet.[134]

134 *HW Neville (Sunblest) v William Press & Son* (1982) 20 BLR 78.

However, that cost cannot be deducted from the final account by the architect in the final certificate, because the carpet is not part of the Works. The employer has the choice either of taking action against the contractor for the cost, or – the simpler method – after having served the appropriate notices under clauses 4.15.3 and 4.15.4, of setting-off the cost against the amount due in the final certificate. It should be noted that when Part 8 of the Local Democracy, Economic Development and Construction Act 2009 commences it will amend the Housing Grants, Construction and Regeneration Act 1996 requirements for notices before payment.

If it was foreseeable that the employer would lay a carpet, but not of the quality actually used or requiring such care in laying, the contractor would be liable only for the kind of costs that would be reasonably foreseeable.[135]

101 The contractor says that, under IC, it has no liability for defects appearing after the end of the rectification period. Is that correct?

The rectification period in all standard building contracts, despite its name, does not signify the maximum period during which the contractor is liable for rectifying defects. It is there for the contractor's benefit. The rectification period in IC is an example. Under the terms of the contract, the contractor's obligation is to construct the building in accordance with the contract documents (clause 1.1), which probably consist of drawings and a specification. If the contractor does not comply with the contract documents, amended if appropriate by architect's instructions, it is in breach of contract.

When the contractor offers the building to the architect as having reached practical completion and the architect has issued a certificate to that effect, the building should have no visible defects and there should be very little work left to complete.[136] The contractor's licence to occupy the site expires at practical completion and it must leave. If there is anything found to be not in accordance with the contract documents and architect's instructions at this point, the contractor is in breach of contract.

If there was no rectification period, the employer would have the right to notify the contractor of the defects, seek competitive

135 *Hadley v Baxendale* (1854) Ex 341.

136 *Westminster Corporation v J Jarvis & Sons* (1970) 7 BLR 64.

quotations from other contractors for making good and then have the defects corrected by the lowest tenderer and recover from the original contractor as damages the total cost of such making good, including professional fees. The employer would have the option to request the contractor to make good the defects at its own cost but, in the absence of a rectification period, the employer would not be bound to do so and the contractor would not be bound to make good although it would be liable for the breaches of contract. The contractor's liability would extend for 6 years from practical completion (12 years if the contract was executed as a deed) in accordance with the Limitation Act 1980.

The rectification period (formerly the 'defects liability period' under previous JCT forms of contract) was introduced to give the contractor the right to return to site and make good any defects notified at the end of the period. It is obviously less costly to the contractor to make good its own defects than to pay the cost involved if other contractors do the work. If the employer does not want the contractor to make good such defects, the architect may issue instructions to that effect to the contractor and an 'appropriate deduction' is to be made from the Contract Sum (clause 2.10). Unless the reason for the instructions concerns some serious fault on the part of the contractor, such as failure to act despite several reminders, the deduction from the Contract Sum can be only what it would have cost the contractor to make good.[137]

It is clear from the contract that the contractor's right to return to site extends only to those defects that appear during the rectification period. Any defects that appear afterwards are still breaches of contract and of course the contractor is still liable for them to the end of the limitation period. The employer is entitled to deal with them as though there was no rectification period as noted above.[138]

102 Can the rectification period be extended to deal with defects discovered and rectified at the end of the period?

This is actually a very good idea, but it is not provided for under the JCT contracts. However, there is provision under GC/Works/

137 *Wiliam Tomkinson and Sons Ltd v The Parochial Church Council of St Michael* (1990) 6 Const LJ 319.

138 *Pearce & High v John P Baxter & Mrs Baxter* [1999] BLR 101.

1 clause 21. This contract refers to the rectification period as the maintenance period. This is a misnomer and a contractor's obligations under this contract are broadly similar to JCT contracts where a rectification period is specified. If GC/Works/1 really did refer to a maintenance period, it would imply that the contractor was obliged to keep the building in pristine condition. Clause 21 divergences from the equivalent clauses in JCT contracts, because clause 21(4) provides that the maintenance period will apply in full to remedial works from the date that they are made good. Therefore, if a defect is notified to the contractor a few days before the end of a 12 months' maintenance period and if the contractor deals with it, a fresh 12 months' period will begin to run for that remedial work. The position could get extremely complicated if there are numerous defects rectified over a lengthy period. Under this contract, the architect has to keep a careful record of the dates defects are notified and made good in order to properly track the new maintenance period. An employer was once heard to say that she intended to keep the whole of the second half of the retention, because she could not be sure that the rectification work was carried out properly and she needed some cash in hand in case she had to fund future remedial work herself. Needless to say, that would be a breach of contract although, in some instances, understandable.

103 The contractor denies liability for a serious defect notified during the rectification period and submits an unsolicited report from an expert which supports its position. Has the report any status under the contract?

A defect in the Works which is due to the work, goods or materials not being in accordance with the contract is a breach of contract on the part of the contractor. Usually, the existence of a defect is a matter of fact which all parties acknowledge. However, on some occasions, the contractor may deny that there is a defect at all, or it may argue that the defect is not its liability either because it is due to a design fault for which the contractor has no liability or for some other reason. The contractor may try to support its position by submitting a report from a third party which says that the contractor is correct. Such a report has no status under the contract in the sense that there is no provision in the contract for such a report to be produced unbidden by the contractor. However, most contracts contain a clause which permits the architect to instruct the contractor to carry out inspections, opening up of work and testing. There is no doubt that an architect may instruct such work which will often generate a report. However, the purpose of such testing is simply to determine the true state of affairs, usually whether or not there is a defect in the work.

It is important to remember that initially it is for the architect to decide whether a defect exists. The architect may decide to instruct opening up or testing and may decide, with the employer's agreement, to commission an expert's report. The architect may consider the report from the contractor's expert, but it is for the architect to decide, possibly in light of all the reports, whether the defect exists and whether to instruct the contractor to comply. If the contractor simply refuses to rectify work and submits a report in support, the architect has the power under most contracts to have the work done by others and then either the cost is deducted from the Contract Sum or the employer will deduct it from any future money owing to the contractor. The precise way in which the cost is recoverable will depend on the precise terms of the contract. At that stage or earlier, the contractor may decide to challenge the architect's view in adjudication or arbitration. During those proceedings, the adjudicator will consider all reports and submissions from both sides before making a decision.

104 What if the employer refuses to allow the architect to carry out the inspection at the end of the rectification period?

Under SBC, clause 2.38 deals with the inspection of the building for defects at the end of the rectification period. It is the architect's obligation to inspect and deliver a schedule of defects to the contractor no later than 14 days after the end of the period. If the architect fails to do that, the defects still amount to breaches of contract on the part of the contractor, but it cannot be compelled to make them good although it is responsible for the cost of doing so. This cost is measured not on the basis of what it would cost the employer, but what it would have cost the contractor.[139] The duty will also be included as part of the architect's services under the terms of engagement.

139 *Pearce & High v John P Baxter & Mrs A Baxter* [1999] BLR 101.

So far as the architect is concerned, it matters not why the employer refuses access. The architect's obligation to inspect is clearly balanced by the employer's obligation to allow entry to every part of the building. They may be particular difficulties, for example the building may be let and the tenants may refuse entry, but that is a matter for the employer, not the architect. If the employer is unsuccessful in securing entry for the architect, there is nothing further the architect can do. The certificate of making good cannot be issued under clause 3.39 and therefore the final certificate cannot be issued. Effectively, the employer has repudiated obligations under the contract and the architect is entitled to accept the repudiation and claim damages (which may be quite slight at this late stage in the project).

There is always the possibility that the employer may instruct the architect not to inspect and the contractor can subsequently receive the certificate of making good. The architect can suggest that course of action, but the architect should not recommend it. In situations like this where the employer is clearly at fault, architects often get into difficulties by failing to properly analyse the position and, instead, they throw themselves into problem-solving mode to attack a difficulty that is not of their making and for the solving of which they are unlikely to receive any thanks. More likely, such architects will eventually be blamed for failing to effect the rectification of defects. In this situation, the plain fact of the matter is that the employer has prevented the architect from carrying out his or her obligations under the contract. The architect is powerless to continue.

If the architect has no option but to accept the employer's conduct as repudiation, it will be for the employer and the contractor to sort out the rest of the contract. This will be bad news for the employer, because, under clause 3.5, the contractor can expect the employer to appoint a replacement architect. In practice, the situation is likely to be resolved by a commercial deal between the employer and the contractor.

105 What if an architect forgets to issue a list of defects at the end of the rectification period under SBC?

This happens more often than might be imagined. If the contractor does not receive the list within 14 days from the end of the rectification period the most obvious thing to do is to issue the list just as soon as the architect remembers. Although the architect is in breach of duty under the contract, there is a good chance that the contractor will simply get on with the remedial work. The contractor may refuse to do the work, arguing that it is under no obligation to attend to any defects which are notified more than 14 days after the end of the defects liability period. The contractor may be under the common, but mistaken, impression that the end of the rectification period marks the end of its liability for defects. Not so. The contractor's obligation is to carry out and complete the Works in accordance with the contract. If it fails to do this, it is in breach of contract as may be evidenced by defects appearing during the rectification period. The contract provides for the contractor to return to site and rectify the defects. If it does not wish to take advantage of this opportunity, the unrectified defect amounts to an unremedied breach of contract. The contractor is not obliged to rectify it, but it is liable for damages for the breach. The employer may engage others to correct the defects and recover the cost of rectifying them. As a result of the architect's failure to issue the list within the contractual timescale, the employer cannot recover the whole of the cost of rectification. Strictly, the employer may only recover what it would have cost the contractor to correct the defects.[140] In practice, the employer would simply issue a withholding notice and deduct the cost from monies due to avoid paying the money and having to sue for its return. It should be noted that when Part 8 of the Local Democracy, Economic Development and Construction Act 2009 commences, it will amend the Housing Grants, Construction and Regeneration Act 1996 requirements for notices before payment.

MW clause 2.10 and MWD clause 2.11 make provision for a rectification period. There is no specific requirement for a schedule of defects, simply for notice to be given by the architect who it seems can require defects to be made good at any time during the period. It is good practice to issue a list and, provided that the architect is not very late in doing so, the contractor probably will have little ground for complaint.

140 *Pearce & High v John P Baxter & Mrs Baxter* [1999] BLR 101.

106 Is there a time limit within which a contractor must remedy all defects notified at the end of the rectification period?

SBC clause 2.38 states that the contractor must make good the defects within a reasonable time. Some other contracts do not include any reference to the matter at all. Where the contract is silent, it is likely that the law will imply a term to the effect that the contractor must rectify the defects within a reasonable time. The question then is the more difficult one of 'what is a reasonable time?' It is easy, and accurate, to say that what is a reasonable time depends on all the circumstances, but it is not very helpful. Ultimately, it is what the adjudicator, the arbitrator or the judge thinks is reasonable in any given situation which is important. The things which will be taken into account are the number of defects and the difficulty of rectification. If rectification requires serious opening up of the fabric, it may well take some time. The urgency of a particular defect will be a factor as will other work in which the contractor is engaged. The fact that a contractor is busy with other work is unlikely to weigh heavily with an adjudicator in view of the fact that each defect is, of course, a breach of contract. Highly specialised work may take longer to organise and rectify, because of the difficulty in securing the appropriate sub-contractor.

It is relatively common for an employer to demand that the contractor fits in with the employer's arrangements when rectifying work. The employer may say that he will only allow access on certain days and perhaps certain hours on those days. Most standard form building contracts say nothing about such things and, while it will be understood that a contractor returning to site will not be allowed a completely free run of the building at such times as it sees fit, it will be implied that the contractor is entitled to access to carry out the remedial work at reasonable times. It may suit both parties for the contractor to do the rectification in small parcels with minimum disruption or, alternatively, to dispose of the whole of the defects in one continuous period of high activity. Either approach could be reasonable, depending on the circumstances of both parties. The one certain thing is that it is to the advantage of both parties to sort out a straightforward timetable without recourse to legal action which may result in an order which suits neither party.

Chapter 15

Valuation and payment

107 MW: Can the contractor insist on agreement on price before carrying out the work?

MW deals with instructions in clause 3.4. Variations and the valuation of such instructions is covered in clause 3.6. Clause 3.6.2 states that the architect and the contractor must endeavour to agree a price before the contractor carries out an instruction, but clause 3.6.3 sets out the architect's power to determine the value of work if no agreement can be reached.

Some contractors have had bad experiences with instructions in that they allege that they comply with an instruction and wait a long time for or never receive payment. Where the amounts are relatively small, it is no use arguing that the contractor has the option of seeking adjudication; the cost would be prohibitive.

The contractor can certainly insist that the architect endeavours to reach an agreement first, but it is impossible to insist that another person agrees anything to which he or she objects. Clauses requiring the architect and the contractor to try to agree something are not much use. Can the contractor argue that the architect did not 'endeavour' to agree a price. Yes, of course, but it will be a well nigh impossible task to prove. The architect need only say that they could not agree on the price. It is as simple as that. The contractor cannot refuse to comply with an instruction because there has been a failure to agree the price. It should not be forgotten that the fact that the contractor has carried out additional work does not entitle the contractor to payment. There must be an instruction properly issued under the terms of the contract. If the instruction is properly issued, the contractor should not be out of pocket.

108 Is the contractor obliged to stick to a low rate in the bill of quantities if the amount of work is substantially increased?

Contractors occasionally insert the wrong rate in bills of quantities. Sometimes it is done on purpose. But even if it can be conclusively demonstrated to be inaccurate, it is of no consequence; the rate or price in the bills must be used as the basis for valuation and it can be adjusted only to take account of the changed conditions and/or quantity. The contractor has contracted on the basis that variations may be ordered in the Works, and the employer has contracted to pay for them on this basis. Neither party can avoid the consequences on the grounds that the price in the bills was too low. The contractor's only hope is that it can be shown that the rate in question is narrow in its application and, therefore, not capable of being applied if the amount increases. The matter has been settled by the courts long ago[141] and revisited more recently with essentially the same result.[142] A contractor will sometimes take a gamble by putting a high rate on an item of which there is a small quantity or a low rate on an item of which there is a large quantity in the expectation that the quantities of the items will be considerably increased or decreased respectively. If the contractor's gamble succeeds, it will make a nice profit. If it fails, the contractor may lose a considerable sum. It is not unlawful, but rather part of a contractor's commercial strategy.[143]

109 SBC With Quantities: The contractor put in a very high rate for an item of which there were only 3 no. in the bills of quantities and it was subsequently found necessary to instruct over 200 no. of these items. Is the quantity surveyor in order in reducing the unit rate?

The answer to this question is virtually the same as the last question. It is the contractor's right to price the bills in any way it chooses. However, the contractor runs the risk that low-priced items may be increased in quantity and high-priced items decreased in quantity. If

141 *Dudley Corporation v Parsons & Morrin Ltd*, 8 April 1959 unreported.

142 *Henry Boot v Alstom Combined Cycles* [1999] BLR 123.

143 *Convent Hospital v Eberlin & Partners* (1988) 14 Con LR 1.

the contractor is lucky and puts a high price on an item that is subsequently varied so that much more of the item is required, the contractor gets a windfall. The quantity surveyor is entitled to reduce the unit rate only by a reasonable percentage to reflect economy of scale, but from the starting point of the contractor's bill rate.

110 Can an architect who discovers that the contractor is making 300 per cent profit on some goods it is contracted to supply under MW do anything about it?

An architect might be quite annoyed to discover that a contractor whose tender has been accepted is making a large profit on some items. Usually, the architect never gets to know the profit margin because, even where the architect is designated as the person to value variations under MW or MWD, the only relevant document will be the priced specification or a schedule of rates.

Occasionally, the architect does get to know the build up of some of the rates and that is when the nasty surprises occur. Generally, it is unreasonable for the architect to get upset if a contractor is making a large profit. It must be remembered that the contractor has won the contract, presumably, on the basis of the lowest overall tender. Therefore, if the profit margin on some items is high, it is likely to be correspondingly low on others. When contractors submit tenders, they effectively take a gamble. They have to pitch their tenders at a level that will give them a reasonable return, but not so high that they lose the project to another tenderer.

Theoretically, after carefully considering the project and the site, each contractor will look for items that are few in number but which can fairly confidently be expected to increase substantially. They will be given high profit margins on the basis that it will not affect the total price very much, but will eventually net a large profit. On the other hand, numerous items that can be expected to be reduced or even omitted altogether can be priced at a low profit margin or even, occasionally, at a loss, because they will have a big effect on the total price but, if omitted, the possible loss will be omitted also. It is a gamble, because the contractor may be wrong about its expectations. This approach has been accepted as normal practice by the courts.[144]

144 *Convent Hospital v Eberlin & Partners* (1988) 14 Con LR 1.

The architect can do nothing about the high profit margin if the tender has been accepted. The priced specification is part of the contract and the architect must have regard to it when pricing variations. Architects becoming too enraged at the thought of the 300 per cent profit should consider whether they would want to do something if they discovered that a contractor was making little or no profit at all.

111 What is the significance of retention being in trust?

Retention is when an amount is withheld from sums otherwise certifiable to the contractor to serve as a safeguard against the possibility of defective work or materials or even failure to carry out the Works by the contractor. It is usually termed a 'retention fund'. The amount retained is usually 3 per cent (or sometimes 5 per cent of smaller value contracts) of the work properly executed by the contractor. It is accumulated by deducting the appropriate percentage from the valuation of work at each interim payment. SBC clause 4.18.1, IC and ICD clause 4.10 (if the employer is not a local authority), DB clause 4.16.1 and ACA clause 16.4 all state that the employer's interest is as trustee. Not all contracts create a trust fund for the retention. The retention fund under the MW and MWD contracts is not held in trust.

Where someone is in a position of trust, that person has a duty to exercise any rights and powers for the benefit of the person for whom the trust was created. Some building contracts such as SBC clause 4.18.1 provide that the employer's interest in the retention monies is fiduciary as the trustee for the contractor. The clauses often stipulate that the employer has no obligation to invest the retention money retained. The legal effect of this wording is doubtful and appears to be contrary to the Trustee Act 1925 and the Trustee Investments Act 1961 which impose a duty on a trustee to invest trust monies in specified investments.

Many standard form contracts provide for the employer to set the money aside in a separate bank account at the request of the contractor. This reflects a requirement of the general law whenever the contract provides that the retention money is to be held in trust. The requirement applies in law whether or not the contractor actually makes a request.[145] The account name should make it very clear that

145 *Wates Construction v Franthom Property* (1991) 53 BLR 23; *Rayack Construction Ltd v Lampeter Meat Co Ltd* (1979) 12 BLR 30.

it is a designated trust account[146] and be very clear as to the identity of the beneficiary. It is good practice, although not prescribed in the contract, for the employer to certify the action to the architect. The purpose of holding the retention in trust is to safeguard the contractor's money in the event of the employer becoming insolvent. Therefore, the obligation to set trust money aside cannot be overcome by deleting such clauses. The court would simply imply a clause to similar effect.

112 Is there a problem for the employer who assists the contractor by making an advance payment?

SBC clause 4.8 provides that the employer may, but is not obliged to, make an advance payment to the contractor. The idea is fairly straightforward. If both the employer and the contractor agree that an advance payment is to be made by the employer to the contractor, the amount agreed and the date for payment must be inserted in the contract particulars together with a schedule setting out the times and amounts for repayment. This is achieved by the architect deducting the monthly amount from the amount to be certified. The RIBA standard certification forms make express provision for the insertion of this repayment sum. A form of bond is bound into the back of the contract and it is used to protect the employer if the contractor fails to repay the amount advanced. The contract has a default position that the bond is required unless expressly stated otherwise. It is difficult to envisage an advance payment situation in which a bond would not be required. The provisions of SBC, IC, ICD and DB are virtually identical. There is no advance payment provision in MW or MWD. The provisions do not apply where the employer is a local authority. It is not clear why that should be the case, especially since many local authorities seek ways to dispose of excess money before the end of the financial year and have to make amendments to the standard form contracts to allow advance payment to be made.

The dangers of advance payment do not appear to be great provided that the employer insists on a bond being taken out. Without a bond, the danger is that, having received the payment, the contractor

146 *Bodill & Sons (Contractors) Ltd v Harmail Singh Mattu* [2007] EWHC 2950 (TCC).

becomes insolvent before the project is finished and the employer loses his money. Where there is a bond and the contractor defaults, the employer simply requests the surety to pay the amount for which the contractor has defaulted. If the contractor becomes insolvent, the employer will continue to receive the repayments, the only difference being that they will have to be requested at the time of each certificate.

113 If work is being done on a daywork basis, can the time claimed be reduced if the quantity surveyor thinks that the contractor has taken too long?

The whole topic of dayworks is the subject of much misconception. Most standard forms of contract provide for dayworks only as an option to be used if the normal valuation mechanism is not appropriate. It brings up the rear in the valuation tables, because work done on a daywork basis generally costs more than work valued in any other way. Quantity surveyors tend to be frustrated by this state of affairs and use their own experience to reduce the time claimed if it appears to them that it is longer than it should be. It is in those last five words that the misconception lies.

If the parties have agreed that payment is to be made on a daywork basis, the quantity surveyor has no right to reduce the hours and other resources on the sheets.[147] That is because they have agreed that the contractor will be paid for the hours spent and the resources used, not for the hours that should have been spent and the resources that ought to have been used. Of course, the proviso is that daywork is the agreed form of payment. A contractor is not entitled to be paid on a daywork basis simply because it submits daywork sheets. A contractor will often submit such sheets, because usually payment on that basis is better than valuation at contract rates.

Often the magic formula 'For record purposes only' is added. However, where dayworks is to be the method of valuation in any particular case, the addition of those words has little practical value and certainly do not prevent the contents of the sheets being used for calculation of payment.[148]

147 *Clusky (trading as Damian Construction) v Chamberlain*, Building Law Monthly, April 1995 p. 6.

148 *Inserco v Honeywell*, 19 April 1996 unreported.

114 Is the contractor entitled to loss of profit if work is omitted?

If the contractor has undertaken under a contract to do a certain amount of work for a stated sum of money, it has the right to do it if it is to be done at all. If the contract provides that the work may be omitted, that allows the architect to instruct that the work is to be omitted. However, it does not permit the work to be given to someone else, because that would not be omitting the work but merely transferring it to another party. Architects sometimes wonder if the problem can be overcome by omitting the work from the contract and not giving the work to another contractor until much later in the contract or even after practical completion has been certified. Such an action is likely to be ineffective before practical completion. Whether it would be effective after practical completion is open to question. The key point might well be the intention of the employer at the time the omission was instructed by the architect.

An American case dealt with a contract that is similar to JCT contracts.[149] The contract provided for the omission of work without invalidating the contract and provided that such omissions should be valued and deducted from the Contract Sum. The American appeal court sensibly held that the word 'omission' meant only work not to be done at all. It did not mean that work could be taken from the contractor and given to another contractor. Two English cases have reached similar conclusions.[150]

The position is very straightforward. If the contractor has contracted to do the work, it has the right to do it, and if the work is given to someone else to do, it is a breach of contract entitling the contractor to damages unless both employer and contractor concurred in the action. The damages are calculated on the principle that the contractor is entitled to be put back in the position, so far as money can do it, as if the contract had been properly performed.

Where work is omitted to give to another contractor, damages usually amounts to giving the contractor the profit it would have earned had it carried out the work. Of course, it may be that the

149 *Gallagher v Hirsch* (1899) NY 45.

150 *Vonlynn Holdings Ltd v Patrick Flaherty Contracts Ltd*, 26 January 1988 unreported; *AMEC Building Ltd v Cadmus Investments Co Ltd* (1997) 13 Const LJ 50.

contractor would not have earned any profit – it may even have made a loss. In these circumstances the contractor may be grateful that the burden of carrying out loss-making work has been removed. It hardly needs saying that the contractor is entitled to loss of profit only if a profit would have been earned.

Because this is damages for a breach of contract and not loss and/or expense, there is no power for the architect under the contract to certify such sum to the contractor and, when it is agreed, it should be paid directly from the employer to the contractor without an architect's certificate.

115 Is the employer entitled to delay payment if bank funding is delayed?

All the standard form contracts have precise provisions which state the way in which payments are to be made to the contractor and the timing of such payments. Typically, the quantity surveyor prepares a valuation and within 7 days the architect issues a certificate for the amount in the valuation or such other amount as the architect decides is due to the contractor. Usually, the employer has 14 days from the date of issue of the certificate in which to pay the amount certified. Where there is an architect's certificate the employer must pay the amount certified within the 14 days unless a written withholding notice has been issued no later than the prescribed period (5 days in JCT contracts, 7 days under the Scheme) before the final date for payment.[151] Where a withholding notice is served, it must be an effective notice and the grounds must be valid grounds.[152] A withholding notice is essentially in respect of what, in litigation, would amount to a set-off or counterclaim. It must be a valid reason for refusing to pay the whole or part of the money due to the contractor. Therefore, if the employer issued a withholding notice which gave as grounds for withholding: 'I have not enough money', that would not be a valid ground. If the employer says that bank funding is delayed, that is tantamount to saying that he or she has no money. So far as

151 *Rupert Morgan Building Services (LLC) Ltd v David Jervis and Harriett Jervis* [2004] BLR 18. Part 8 of the Local Democracy, Economic Development and construction Act 2009 amends the notices requirements, but at the time of writing no commencement date had been fixed.

152 *Windglass Windows Ltd v Capital Skyline Construction Ltd and Another* (2009) 126 Con LR 118.

the contractor is concerned there is a binding contract which provides that money must be paid. If the employer fails to pay, that is a breach of contract. Therefore, the simple straightforward answer to this question is 'No'.

116 Is it true that a change in the scope of work can result in a re-rating of the entire bill of quantities?

A contractor who has under-priced a project or who has been on the receiving end of many architect's instructions often contends that the effect of all the instructions is to turn the project into something entirely different from what the contractor priced in its tender. The phrase used is usually that the instructions 'changed the whole scope and character of the work'. If the variations have had this effect, the contractor would be right in its assertion despite clause 3.14.5. Everything depends on what the parties expressed as their intentions in the contract. One of the important reasons for the variation clause in a contract is to prevent the contract being put at an end by an instruction to the contractor to alter or modify the Works in some way. SBC clause 3.14.5 expressly so states. If there was no variation clause, an alteration to the Works would necessitate agreement by both parties and perhaps a renegotiation of the contract.

The usual example is, if the contract was to build one house and the architect issued a variation to add another similar house, it would probably vitiate the original contract because the scope of the work would have been doubled and the contract would be markedly different from the contract which the contractor undertook to carry out. The contract may be frustrated. If, however, the contract was to build one hundred houses and the architect issued a variation to add one house, it would be unlikely to vitiate the contract because the scope would have been increased by only 1 per cent; it would be the same contract with a minor variation to the Works. Obviously an architect's instruction, or a series of such instructions, which had the effect of altering an office into a factory would vitiate the contract.

Therefore, the answer to this question is 'Yes in principle'. However, in practice the situation will rarely arise. Certainly, even where the architect issues a substantial number of variations, a contractor will find it difficult to argue that the whole scope of the Works has been changed. In *McAlpine Humberoak v McDermott*

International Inc (No. 1)[153] it was held at the first trial that the contract was frustrated because of the large numbers of drawings which were issued. The Court of Appeal held that it was not frustrated, because there was provision for the variations to be properly valued and for the contractor to be recompensed for the delays. Even where a housing contract in the sum of £126,000 was subsequently amended by variation to become a contract of £1.45 Million, the court seems to have been unconcerned.[154] Admittedly, there appears to have been some unusual factors in that instance, but it is thought unlikely that a contractor's claim that the whole scope of the Works has been changed will usually stand much chance of success.

117 German light fittings were specified. Can the contractor claim extra money because the exchange rate has altered to its detriment?

Some contracts, particularly those used for international work, make express provision for the situation which may arise where goods from a country, other than the country in which the work is to be carried out, are specified. However, JCT contracts make no such provision. When a contractor is invited to tender, it can decide not to do so, because the risks are unacceptable. If a contractor does decide to tender, it is taking on board a number of risks, for example, the risk that obtaining adequately skilled labour will be difficult. The JCT standard contracts have fluctuations provisions for increases in labour and materials, but only if such clauses are said to apply. In this instance, it is not an increase in the price which is the problem. The problem is that the price remains the same but the monetary exchange rate has changed, resulting in a potential loss for the contractor. This is one of the risks which the contractor must allow for in its price. Therefore, unless the price has actually increased and the relevant fluctuation clause is operative, the contractor has no claim against the employer for more money. The situation would be different if the contractor can cite one of the relevant matters under SBC clause 4.24 and realistically argue that the relevant matter caused a delay which prevented the contractor from placing an order in time to avoid the exchange rate alteration. Assuming that the contractor

153 (1992) 8 Const LJ 383.

154 *Bruno Zornow (Builders) Ltd v Beechcroft Developments Ltd* (1990) 6 Const LJ 132.

could marshal enough supporting evidence, it would be entitled to the extra cost as part of its application for loss and/or expense.

118 Under what circumstances is the contractor entitled to the costs of acceleration?

A case decided in 2000, defined 'acceleration' like this:

> 'Acceleration' tends to be bandied about as if it were a term of art with a precise technical meaning, but I have found nothing to persuade me that that is the case. The root concept behind the metaphor is no doubt that of increasing speed and therefore, in the context of a construction contract, of finishing earlier. On that basis 'accelerative measures' are steps taken, it is assumed at increased expense, with a view to achieving that end. If the other party is to be charged with that expense, however, that description gives no reason, so far, for such a charge. At least two further questions are relevant to any such issue. The first, implicit in the description itself, is 'earlier than what?'. The second asks by whose decision the relevant steps were taken.
>
> The answer to the first question will characteristically be either 'earlier than the contractual date' or 'earlier than the (delayed) date which will be achieved without the accelerative measures'. In the latter category there may be further questions as to responsibility for the delay and as to whether it confers entitlement to an extension of time. The answer to the second question may clearly be decisive, especially in the common case of contractual provisions for additional payment for variations, but it is closely linked with the first; acceleration not required to meet a contractor's existing obligations is likely to be the result of an instruction from the employer for which the latter must pay, whereas pressure from the employer to make good delay caused by the contractor's own fault is unlikely to be so construed.[155]

Unless expressly so stated in the building contract, the architect has no powers to instruct the contractor to accelerate work. The contractor's obligation is to complete the work within the time specified or, where no particular contract period is specified, within a

155 *Ascon Contracting Ltd v Alfred McAlpine Construction Isle of Man Ltd* (2000) 16 Const LJ 316.

reasonable time. The employer cannot insist that the contractor completes earlier than the agreed date in the absence of an express contract term.

No JCT traditional contracts give either the architect or the employer power to order the contractor to accelerate. There is, however, such a power in the ACA 3 form, clause 11.8. The contractor should be able to obtain payment where the architect orders acceleration of the work under a term of the contract or the employer and the contractor agree acceleration.

A contractor will sometimes base its case on the architect's failure to give an extension of time. The contractor will often put more resources into a project than originally envisaged and then attempt to recover the value on the basis that there was no realistic alternative, because the architect failed to make an extension of the contract period. A contractor in this situation contends that, as a direct result of the architect's breach, it was obliged to devote more resources to the project so as to finish by the date for completion, otherwise there was a danger that the employer would levy liquidated damages. This claim tends to be advanced whether or not completion on the due date is actually achieved.

Before this argument can be entertained, the key question is: 'What was the true cause of the acceleration?' The contractor's difficulty is that if the architect wrongfully fails to make an extension of time, either at all or of sufficient length, the contractor's redress under the contract is adjudication or arbitration. If the contractor is entitled to an extension of time, it should simply continue the work, knowing that it will be able to recover its prolongation loss and/or expense, and any liquidated damages deducted, by referring the dispute to adjudication or arbitration. The true cause of the contractor's acceleration is not any breach by the architect, but simply a decision by the contractor to put in more resources. Of course, a contractor in this position may not be entirely confident that adjudication or arbitration will result in reimbursement of money lost. There are few certainties and the liquidated damages may be high. Few would pretend that justice will inevitably be done in adjudication, arbitration or legal proceedings. The contractor may consider that it is less expensive to accelerate rather than face liquidated damages with no guarantee that an extension of time will ultimately be made, even without recovering acceleration costs. It may simply be a commercial decision for the contractor. It is thought that a claim of this kind has little prospect of a successful outcome.

A common situation is where a contractor accelerates without any agreement with the employer or instruction from the architect. The result may be that some time is recovered and an extension of time may be avoided. In this situation, a contractor may argue that, had it not accelerated, there would have been a delay to completion. Using a computer model, it may be demonstrated that the completion date would have been exceeded had the contractor not accelerated. Notwithstanding that, in most such cases the contractor will not find it easy to argue that it was doing other than using best endeavours to reduce delay, and there is no clause in the JCT traditional contracts that could be used to reimburse a contractor in this position.

Although acceleration has been considered in another case, the conclusion was so bizarre as to render it extremely suspect.[156] In this case, the court decided that the contractor was entitled to recover the cost of acceleration if an extension of time was justified, but refused, and the liquidated damages were 'significant'. So far so good, albeit somewhat off the mainstream view. However, the court held that the contractor was entitled not only to the cost of acceleration but also to loss and/or expense for the prolongation that would, but for the acceleration, have taken place. On any view, that amounts to double recovery. To summarise the position:

- There is no clause in traditional JCT contracts and nothing which the general law would imply that gives the architect power to instruct the contractor to accelerate.
- The contractor and the employer can enter into a separate agreement to accelerate, but payment cannot be made under the contract.
- A contractor that accelerates without an agreement from the employer cannot recover the costs of doing so except in wholly exceptional circumstances.

119 What is the effect of agreeing payment 'in full and final settlement'?

It is relatively common for payment to be offered 'in full and final settlement'. Great care must be taken when faced with these words. The

156 *Motherwell Bridge Construction Ltd v Micafil Vakuumtechnik and Another* (2002) 81 Con LR 44.

law is quite complex and based on what is known as 'accord and satisfaction'. This is defined as: 'The purchase of a release from an obligation whether arising under contract or tort by means of any valuable consideration, not being the actual performance of the obligation itself. The accord is the agreement by which the obligation is discharged. The satisfaction is the consideration which makes the agreement operative.'[157] If there is accord and satisfaction, it acts as a bar to any action.

If a person agrees to accept part payment and to release the other from payment of the balance, this will be valid if the agreement is supported by fresh consideration or if the agreement is a deed that requires no consideration. The key point is that the creditor must accept something different from the legal entitlement.[158] The law does not accept that a debt can be discharged simply by payment of a lesser sum. Therefore, if a party is owed £500 and the debtor offers £200 'in full and final settlement' of the debt, the creditor is entitled to take the £200 and subsequently take action to recover the balance.

The payment would be validly made to settle the debt if it were made in a different way or, perhaps, in a different place. So that, if £500 is owed, payment of £200 worth of grass seed could represent true accord and satisfaction.

Sometimes a cheque is sent on the basis that payment into the other's bank account will signify acceptance 'in full and final settlement'. If the cheque is simply paid into the account, it is likely that a court would deem that it was accepted on the basis it was paid. It is understandable that a party owed a substantial sum with a cheque in its hand will be keen to recover as much as possible and, therefore, will be anxious to bank the cheque. The answer is to write to the sender noting that the cheque is accepted and will be paid into the bank, not in full and final settlement, but as a part payment of money owing. Then, the cheque should be paid into the account a couple of days later. That allows the sender to stop the cheque if it feels so inclined. Perhaps surprisingly, few cheques appear to be stopped in this situation, the sender preferring to rely on the now useless argument that payment is as indicated by the sender's terms despite the note from the receiving party to the contrary.

157 *British Russian Gazette & Trade Outlook Ltd v Associated Newspapers Ltd* [1933] 2 KB 616.

158 *Pinnel's Case* (1602) 5 Co Rep 117a.

In *Stour Valley Builders v Stuart*,[159] the builders sent an invoice for work undertaken. The Stuarts disputed the amount and sent a cheque for a lesser sum 'in full and final settlement'. The builders cashed the cheque, but telephoned the Stuarts saying that it was not accepted in full and final settlement. The Court of Appeal held that the cheque was not accepted in full and final settlement and, therefore, the builders were entitled to recover the balance. The court said: 'If the creditor at the very moment of paying in the cheque makes clear that he is not assenting to the condition imposed by the debtor, how can it be said that, objectively, he has accepted the debtor's offer?'

However, that comment applies to a situation where the debt is indisputable. If there is a genuine dispute during which one party says it is owed £500 and the other argues that it owes nothing, an offer in full and final settlement by the alleged debtor of £200 which is accepted by the creditor will not enable the creditor to return later for the balance, because accord was reached in settlement of a dispute and the courts encourage parties to settle their differences by agreement. It is crucial to decide whether there is a genuine dispute or whether one party simply does not want to pay. In these situations, legal advice is always necessary.

120 Under DB, the Employer's Requirements asked for special acoustic windows which the Contractor's Proposals did not include. The contract is signed. Can the employer insist on the special windows at no extra cost?

At the root of this question is the priority of documents. In DB, two situations are envisaged: a discrepancy within the Employer's Requirements and a discrepancy within the Contractor's Proposals. In each case, employer and contractor share the duty of informing the other if either discovers a discrepancy. Under clause 2.14.2, a discrepancy in the Employer's Requirements is dealt with in whatever manner is stated in the Contractor's Proposals or, if not so stated, as suggested by the contractor, which the employer can either accept or reject in favour of its own solution. Either way, it is to be treated as a variation (which is the term for a change). A discrepancy in the

159 [2003] TCLR 8.

Contractor's Proposals is covered by clause 2.14.1. The contractor must suggest an amendment and the employer may choose between the discrepant items or the suggestion at no additional cost.

What happens when there is a discrepancy between the Employer's Requirements and the Contractor's Proposals? In this case, the Employer's Requirements asked for special acoustic windows, but the Contractor's Proposals, by accident or design, does not include them. The contract does not expressly address this problem. Footnote [3] emphasises the importance of removing all discrepancies between the two documents. Unfortunately, discrepancies will occur. The usual way of resolving such matters is on the basis of priority of documents.

It is often mistakenly said that the third recital of the contract covers the position and shows that the Contractor's Proposals take precedence. This recital provides that the employer has examined the Contractor's Proposals and, subject to the conditions, is satisfied that they appear to meet the Employer's Requirements. Whatever else may be said about this recital, the use of the word 'appear' and the fact that it is subject to the conditions is significant. Without these, the employer is satisfied that the Contractor's Proposals meet the Employer's Requirements. The addition of 'appear' makes clear that the satisfaction is simply dealing with surface appearance. One might say 'on the face of things' or, as the lawyers used to say before Latin became unfashionable, *prima facie*. The dictionary defines 'to appear' as 'to give an impression'. It is clearly not intended that, under the contract, the employer or his or her advisers are intended exhaustively to check the Contractor's Proposals to *ensure* that they meet the Employer's Requirements. Had such a thing been intended, it would have been easy for the draftsman to have used clear words to that effect. If the employer requested a five-storey office block in the Requirements, the third recital merely records that the employer believes that is what the Proposals provide. That the statement is made subject to the conditions, very clearly tells the reader that the printed conditions have something important to say about the situation.

The wording strongly points to the intention that the Contractor's Proposals will be drafted to meet the Employer's Requirements. In doing so it is merely confirming the philosophy of the contract as can be discerned from the Recitals as a whole. The Contractor's Proposals should be an indication of how the contractor is to comply with the Employer's Requirements – not an indication of how the contractor wishes to construct the project or allocate risk. The wording of the first and second Recitals reflects this.

However, it is misguided to place such reliance on the third Recital, because the role of the Recitals in interpreting a contract is limited. Where the words in the operative part of a contract are clear, the Recitals do not vary that meaning. It is only when the rest of the contract is ambiguous that one turns to the Recitals for assistance. In this instance the contract is clear, as can be seen below. Therefore, the third recital has no, or limited, relevance to this particular question.

The contract is clearly written with the intention that the Employer's Requirements prevail in the following way:

- Clause 1.3 provides that nothing in the Employer's Requirements or the Contractor's Proposals can override or modify the printed form.
- Clause 2.2 provides that the Employer's Requirements prevail over the Contractor's Proposals where workmanship or materials are concerned. Clause 2.2 states, in part: 'All materials and goods for the Works shall . . . be of the kinds and standards described in the Employer's Requirements, or, if not there specifically described, in the Contractor's Proposals . . .'. From that it is clear that it is only if the Employer's Requirements make no mention of the materials and goods that the contractor can turn to the Proposals. Clause 2.2.2 is in very similar words in respect of workmanship.
- Under the terms of the contract, the employer cannot issue a change instructing the contractor to vary the Contractor's Proposals. Clause 5.1 provides that a change means a change in the Employer's Requirements. Nor can the employer instruct the expenditure of a provisional sum in the Contractor's Proposals (see clause 5.2.3). If the Contractor's Proposals prevailed over the Employer's Requirements, it would prevent the employer from issuing changes in respect of the discrepant parts of those Contractor's Proposals. That cannot be what the contract intended. Such changes go beyond matters of design and construction and embrace sequence of work and access, etc.
- The intention of the contract is that the Employer's Requirements and the Contractor's Proposals should dovetail together. Where they do not do so, it would be perverse to

permit the Proposals to take precedence, because the employer is entitled to assume that the contractor is complying with the Requirements.

The answer to the question is clearly that the employer can insist on the special windows, because, where there is a conflict, the Employer's Requirements prevail over the Contractor's Proposals.

Chapter 16

Certificates

121 SBC: Is the contractor entitled to suspend work under the Construction Act if the architect has under-certified?

The right to suspend performance of obligations under the contract is contained in section 112 of the Housing Grants, Construction and Regeneration Act 1996. Section 112(1) states:

> 112(1) Where a sum due under a construction contract is not paid in full by the final date for payment and no effective notice to withhold payment has been given, the person to whom the sum is due has the right (without prejudice to any other right or remedy) to suspend performance of his obligations under the contract to the party by whom payment ought to have been made ('the party in default').

When Part 8 of the Local Democracy, Economic Development and Construction Act 2009 commences, it will amend this section to refer to the suspension of performance of '*any or all* of his obligations'. Further sub-sections proceed to stipulate that at least 7 days' written notice must be given, that the right to suspend comes to an end when payment in full has been made and that the person suspending has, in effect, the right to an extension of any relevant contract period. The 2009 Act will extend the right to include the right of the suspending party to be paid costs and expenses reasonably incurred.

Under SBC, the architect is required to issue interim certificates under clause 4.9. Clause 4.14 essentially repeats the substance of section 112. The final date for payment is stipulated by clause 4.13.1 to be 14 days after the date of issue of the architect's certificate. Therefore, it is clear that there can be no final date for payment unless the architect issues a certificate. There is authority to say that the employer may well be liable if the architect does not properly comply with his or her duties under the contract, including the duty to certify at the intervals prescribed in the contract.[160] The employer's liability would depend on the employer knowing that, first, the architect had such a duty and, second, that the architect was in breach of the duty.[161]

However, here we are not considering a situation where the architect fails to certify at all, but where the architect certifies a lesser sum than the contractor thinks is due. Therefore, the architect has not failed to carry out the duty to certify. Clause 4.13.5 obliges the employer to pay the contractor the amount stated on a certificate (obviously, this is subject to the right of set-off and notification in clauses 4.13.3 and 4.13.4). Therefore, if the employer pays the amount on an architect's certificate, even if that certificate is seriously undervalued, the employer cannot be in breach of contract.[162] The contractor's right to suspend arises only if the amount due under the contract, in this instance it is the sum certified, remains unpaid after the final date for payment.

Although the architect's failure to certify the proper amount may be a breach of contract on the part of the employer, depending on whether the employer was aware of any under-certification, it is clearly not something for which the contractor can suspend. It is worth noting that, although the question is couched in terms of suspending work, both section 112 of the Act and SBC go much further and refer to suspension 'of performance'. In other words, the contractor is entitled to suspend anything at all that the contract requires it to do. The fact that the contract requires the contractor to insure the Works and other matters immediately springs to mind. If the contractor not only suspends work but also suspends all its insurances relating to the Works, the employer will be in a very difficult position.

122 Can an architect issue a negative certificate?

Usually, by a 'negative certificate' what is being referred to is a certificate that shows a negative amount owing from the employer to

160 *Perini Corporation v Commonwealth of Australia* (1969) 12 BLR 82.

161 *Penwith District Council v V P Developments Ltd* (2005) 102 Con LR 117.

162 *Lubenham Fidelities v South Pembrokeshire District Council* (1986) 6 Con LR 85.

the contractor – in other words, a certificate indicating that the contractor has already been paid too much. There are three questions that arise from that:

1 Are there any occasions when an architect may issue such a certificate?
2 If yes, is the contractor then obliged to pay the negative amount to the employer?
3 If yes, do the provisions about notices, particularly withholding notices, work in reverse?

If one looks at the JCT Standard Building Contract (SBC) there is nothing that states that the architect may issue a negative interim certificate. On the other hand, there is nothing to say that the architect may not issue one.

Interim certificates are dealt with under clause 4. Clause 4.9.1 states that the architect must issue interim certificates 'stating the amount due to the Contractor . . .'. Clause 4.9.2 states that interim certificates are to be issued at the periods stated in the contract particulars and after practical completion 'as and when further amounts are ascertained as payable to the Contractor by the Employer . . .'. There seems at first sight to be nothing that entitles the architect to issue a negative interim certificate. Indeed, everything points to certificates stating payments due to the contractor only. The provisions for the issue of the final certificate are the only ones which recognise that there may be a payment due to the employer.

However, clause 4.10, referring to the amount stated as due in an interim certificate, provides that it must be the gross valuation less certain other permitted deductions and the amount previously certified. It is clear, therefore, that if the amount previously certified and the permitted deductions are together more than the gross valuation, any certificate then issued would be showing a negative amount. In practice, this situation can easily arise if a previous certificate is overvalued by more than the total of the work done between the issue of the previous certificate and the new certificate so that the contractor has not carried out work to the value of the overvaluation in the intervening period.

Even in this situation, the standard certification forms issued by RIBA Publishing are, quite rightly, not worded so as to allow the architect to require payment of the balance by the contractor, and an architect who issues the certificate in the form of a letter is not entitled by the contract to word it in any other way. The inescapable conclusion is that the architect may issue a negative certificate, because that is the result of applying the calculation set out in the contract. However, there is no provision for the architect to certify a payment from the contractor to the employer. This is perfectly sensible and in line with the general intention of the contract. Certification is not to provide the contractor with an exact figure. Its purpose is to provide the contractor with cashflow; sometimes the certificate will be slightly less and sometimes slightly more than the amount of work actually carried out.[163]

In the light of those conclusions, question 3 above does not require an answer. In fact, the provisions regarding notices, particularly with regard to withholding, are not written so as to work in reverse. They expressly refer to notices to be issued by the employer and to the contractor. Of course, the notices are included in the contract as a result of the Housing Grants, Construction and Regeneration Act 1996. (It should be noted that Part 8 of the Local Democracy, Economic Development and Construction Act 2009 amends the requirements for notices before payment, but at the time of writing no date has been set for the Act's commencement.) Section 110 of the 1996 Act refers to the giving of a notice by a party not later than 5 days after 'a payment becomes due from him under the contract . . .'. It has been noted above that there is no payment due to the employer from the contractor under this contract. Section 111 simply states that a party may not withhold payment unless a withholding notice has been given. Applying that to the contract, the contractor would not be entitled to withhold payment unless an effective withholding notice was served. But the contractor would have no need to issue such a notice unless there was a contractual obligation that the contractor should pay in the first instance.

Therefore, the answer to questions 1, 2 and 3 appear to be 'Yes', 'No', and 'Not applicable, but No in any event'.

So far as the final certificate is concerned, clause 4.15.2 provides that the final certificate must state an amount due to the employer or to the contractor as the case may be. However, provisions for notices in clauses 4.15.3 and 4.15.4 refer only to the employer. Nevertheless, if in the final certificate the contractor was found to owe money to the employer and the contractor wished to withhold some or all of

163 *Sutcliffe v Chippendale & Edmondson* (1971) 18 BLR 149.

that money, it appears that notices would have to be given under sections 110 and 111 of the 1996 Act.

123 Can an architect who has under-certified withdraw the certificate and issue a revised certificate or simply issue another certificate for the additional money?

The straight answer to this is that most contracts make provision for interim certificates to be issued at stated intervals, commonly monthly. The architect only has the powers conferred by the provisions of the contract and, therefore, additional certificates may only be issued at the intervals stated. For example, if the contract states that certificates may be issued every month, the architect has no power to issue a certificate before the month has elapsed. In practice, the parties can jointly authorise the architect to issue an additional certificate within a shorter period and, in the case of a serious shortfall, no doubt that would be the answer. An architect who has under-certified may not be anxious to make that fact known to the employer and may seek to withdraw one certificate and replace it with another. An architect who has over-certified will also wish to withdraw that particular certificate.

There seems to be little doubt that, based on well-known principles applicable to arbitrators' and adjudicators' awards and decisions, the architect will be able to withdraw the certificate and issue a new certificate if the reason for the withdrawal amounts to a simple arithmetical error or some other similar accidental slip. However, that would not cover a situation where an architect has wrongly certified for other reasons. There appears to be no judicial decision directly on this point. The problem is that, in certifying, the architect is exercising professional judgment and the certificate is the formal expression of that judgment. Therefore, in issuing the certificate, the architect is expected to have used reasonable skill and care in coming to that judgment. The withdrawal of the certificate and its substitution with another may suggest that the architect has been negligent in the issue of the first certificate. In practical terms, the withdrawal of a certificate and the substitution of one for a greater amount, will draw no complaints from the contractor, but possibly many serious questions from the employer. Replacing a certificate with one of less value may, depending on the value, cause the contractor to seek adjudication and may still draw questions from the employer. Although there may be some advice to the effect that an architect can withdraw and replace any certificate, it is thought that the better view is that, once issued and with the exception of accidental errors previously noted, a certificate cannot be withdrawn by the architect and can only be overturned by agreement of the parties or by an adjudicator, an arbitrator or a court.

124 What is the payment position if the architect refuses or fails to check and sign daywork sheets?

Standard form contracts do not prescribe dayworks as the normal method of valuing and paying for work carried out. Valuation by dayworks is usually confined to situations where additional or substituted work cannot be valued by measurement. Occasionally, an entire contract may be let on the basis of daywork, but that is the exception and usually because the full extent and nature of the work is not clear at the outset. This may be a satisfactory method of valuation for the contractor, because it ensures that it will, at least, recover the cost of doing the work plus percentages to cover supervision, overheads and profit; how it is not a satisfactory method so far as the employer is concerned, because it gives no incentive to the contractor to work efficiently. It should be noted that the quantity surveyor has no right to vary the hours and other resources on the sheets.[164]

SBC makes provision for work to be valued on a daywork basis in clause 5.7, but only where the varied work cannot be valued in any other way. It is essentially a last resort. In order to work properly, the quantity surveyor should notify the contractor in advance of the intention to value using daywork. Sometimes that cannot be done, because the quantity surveyor may not know, until the work has been carried out, that daywork is the only sensible option. Often, contractors will complete and submit daywork sheets as a matter of course for all the work carried out. Daywork is usually most lucrative, because the contractor is paid for all the hours spent on the work. Of course, the quantity surveyor is still under no obligation to accept daywork as the method of valuation if the work can properly be measured.

164 *Clusky (trading as Damian Construction) v Chamberlain, Building Law Monthly*, April 1995, p. 6.

Clause 5.7 provides that daywork sheets (referred to in the contract as 'vouchers') must be verified by the architect if delivered by the contractor not later than the end of the week following the week in which the work was carried out. Verification is normally carried out by signing the sheets. Where in the contract, such as in SBC clause 5.7, there is a system of verification, but the architect has failed or refused to comply, the sheets will be evidence of the work done unless they can be shown to be inaccurate or that the contractor did not comply with the system.[165] The quantity surveyor then must use the unsigned sheets as the basis for any payment which is to be based on daywork.

125 If the contractor is falling behind programme, is the architect justified in reducing the amount of preliminaries costs in interim certificates?

The preliminaries section of the bills of quantities or specification should be priced by the contractor when submitting a tender. It may price every item individually based on its anticipated costs for that item or it may simply allow a percentage against preliminaries which is calculated on the cost of the measured work. Sometimes a contractor will simply insert a lump sum figure as a total for all the preliminaries items without giving any hint of the way in which it has been calculated. It is usual, when preparing an interim valuation, to include a sum to represent a reasonable proportion of the contractor's preliminaries price. If the contractor has priced individual items, then this sum can be calculated using those items. If the preliminaries figure is a lump sum, the valuer may simply divide the sum by the number of months in the contract period to represent monthly valuations.

If the contractor falls behind programme so that it seems likely that the total contract period will be exceeded, it is common for the monthly preliminaries amount to be reduced in order to extend the total, unchanging, preliminaries amount over the longer contract period. This is sometimes referred to as 'adjustment of preliminaries'. It is always something of a guess, because when the decision to adjust or extend the preliminaries is made, it will not be known with

165 *JDM Accord Ltd v Secretary of State for the Environment, Food and Rural Affairs* (2004) 93 Con LR 133.

any precision how much the overrun will be. The justification for reducing the preliminaries sum in individual valuations, so that they are spread out over a longer period, is that the preliminaries costs should be related to the actual work carried out. Therefore, if the contractor has only carried out two thirds of the work, it should be entitled only to two thirds of the preliminaries. The contractor's argument is likely to be that it is suffering the same losses each month whether or not the total period is prolonged. Against this it can be rightly said that if the contractor had not caused the delay, it would not be suffering the losses.

Depending upon the terms of the building contract, the position will be different if the contractor is entitled to loss and/or expense for the overrun period. If the whole of the prolonged period is caused by reasons which entitle the contractor to loss and/or expense, there will be no adjustment of preliminaries at all. The contractor will be entitled to the full amount of the preliminaries each month (e.g. one tenth of the total preliminaries for each month of a 10 months' contract). The loss and/or expense will deal with the additional period. In practice, the situation will probably be that responsibility for the overrun will probably be partly the fault of the contractor and partly for reasons which entitle it to loss and/or expense. In such a case, the extension of preliminaries will be carried out for the contractor default period and loss and/or expense will be ascertained, on the contractor's properly submitted application, by reference to actual loss and actual expense.

126 Must the architect certify the amount in the quantity surveyor's valuation?

SBC clause 4.11 states that interim valuations must be made by the quantity surveyor whenever the architect considers them to be necessary for ascertaining the amount to be certified. The exception to that is if fluctuations option C, formula adjustment, is applicable, in which case the valuation must be made before each interim certificate is issued. Therefore, it is clear that, in general, the architect need not ask the quantity surveyor to prepare a valuation. However, the question is whether the architect, having requested and received the quantity surveyor's valuation, must certify the amount the quantity surveyor has indicated or whether the architect is entitled to certify a different amount. It is thought that the architect is entitled to certify a different amount for the following reasons:

- Even though the contract provides for a valuation to be carried out if the architect believes it to be necessary, it is the architect who is charged with certifying the amount due to the contractor. If the architect had no power to vary the amount valued by the quantity surveyor there would be little or no point in an architect's certificate being necessary before payment is due. The contract could easily provide for the employer to pay on the basis of the quantity surveyor's valuation.
- The architect is not obliged to seek a valuation before each certificate and on those occasions when a valuation is not sought, the certificate will not be based on a valuation, but on the architect's own opinion. If the architect is entitled to use his or her discretion whether to ask for a valuation and, if not, to determine the amount to be certified, it should logically follow that the architect is entitled to use discretion even if a valuation has been requested.
- The architect has a duty to be reasonably satisfied with the correctness of the quantity surveyor's valuation and to make the final decision on the amount to be certified.[166] That does not mean that the architect is obliged to check the whole of the valuation, still less to carry out a separate valuation. However, the prudent architect will request the quantity surveyor to provide a general breakdown so that the architect can check that defective work has not been included or that some large items have been omitted.

Something which is very common, but which an architect should never do, is to simply copy down all the figures from the quantity surveyor's valuation on to the interim certificate without giving any thought to whether they are correct. Such action is bordering on, if not actually, negligence.

127 Under IC, if the time for issuing a withholding notice has expired, but some serious defects come to light, can the employer set off the value against the amount certified?

This is the employer's worst nightmare. Under clause 4.8.2, not later than 5 days after the date of issue of an interim certificate, the

166 *Sutcliffe v Thackrah* [1974] 1 All ER 859; *RB Burden Ltd v Swansea Corporation* [1957] 3 All ER 243; *Cantrell and Another v Wright & Fuller Ltd* (2003) 91 Con LR 97.

employer must give a notice to the contractor stating the amount the employer proposes to pay, to what it relates, and the basis on which it is calculated. Under clause 4.8.3, not later than 5 days before the final date for payment of an interim certificate, the employer may give a written notice to the contractor stating any amount or amounts proposed to be withheld and the ground or grounds for the withholding. If the notice is not given, clause 4.8.4 makes clear that the employer must pay the amount stated in the clause 4.8.2 notice. If the employer has failed to give a clause 4.8.2 notice, the amount to be paid is the amount stated as due in the certificate.

The scheme of notices is straightforward. The first notice is the employer's opportunity to tell the contractor that the employer disagrees with the architect's certificate. Provided a proper calculation of the money the employer considers is due is given, that is all the employer need pay. If the employer does not give the first notice, it is assumed that the certified amount is correct and the employer's only chance to avoid paying it is to give the second notice with adequate figures and reasons showing why part or all of it is to be withheld. The deadline for the second notice is 5 days before the final date for payment (which is 14 days from the date of issue of the certificate). If serious defects make their appearance after the deadline, the employer has no option but to pay. If the employer fails to pay, there will be no viable defence if the contractor goes to immediate adjudication. There are similar provisions relating to the final certificate in clause 4.14.

It has been known for an employer who can put before an adjudicator positive evidence about the existence and value of the defects to persuade the adjudicator to support the withholding even though it was made without proper notice but, to be frank, that depends upon the appointment of an adjudicator with an inadequate understanding of his or her role.

Realistically, the employer must pay and, if the defects are not corrected by the date of the next valuation, the architect must certify the amount properly due taking the defects into account. This may result in a negative certificate. In most cases, the defects will be made good and the overpayment to the contractor will rectify itself as work proceeds. In rare cases, a dispute may develop and the contractor may become insolvent, leaving the employer in the position of having overpaid. In that situation, the employer may well look to see whether some action is possible against the certifying architect on the basis that if the architect had properly carried out inspection duties, the serious defects would have been discovered earlier. Whether that approach would be successful depends on the circumstances of each individual case. It should be noted that Part 8 of the Local Democracy, Economic Development and Construction Act 2009 amends the notice provisions in the Housing Grants, Construction and Regeneration Act 1996 which are reflected in the above clauses, but at the time of writing no commencement date for Part 8 has been fixed.

128 If the employer and the contractor agree the final account, should the architect issue a final certificate in that amount?

All the standard form contracts require certificates to be issued by the person named in the contract as the architect or the contract administrator. A certificate is the formal expression of the architect's professional opinion.[167] In short, it is a very serious document and not something to be issued without careful thought.

It is quite common for the employer and the contractor to effectively 'do a deal' at the end of a project and agree between them the amount the employer will pay to close the contract. Such an agreement is often based on the age-old principle of a figure more than the employer really wants to pay and less than the contractor expects. A settlement is sometimes said to be successful when both parties are dissatisfied with it.

In the normal course of events, the issue of the final certificate under any of the standard forms will be the culmination of a process that has been continuing from the commencement of work on site. During this time, the Contract Sum is constantly adjusted to take account of variations and any other matters that the particular contract allows to change the Contract Sum. After practical completion of the Works, if the contractor wishes to submit any further information to the architect (or to the quantity surveyor if the contract stipulates that the quantity surveyor is to value), there is a specific time within which this may be done, usually 6 months. The quantity surveyor completes the adjustment of the Contract Sum and, after consultation with the architect, sends this figure to the contractor. Within a contract-stipulated timescale, the architect issues the final

167 *Token Construction Co Ltd v Charlton Estates Ltd* (1973) 1 BLR 48.

certificate. This certifies the amount that is due to the contractor and that the amount has been calculated in accordance with the terms of the contract.

Obviously, where a settlement figure has been agreed between the parties to the contract, it has not been calculated in accordance with the terms of the contract. Therefore, the architect cannot certify that it is the amount which is objectively due to the contractor. It follows that if the parties agree the amount payable from employer to contractor (usually) to settle the contract, the architect cannot issue a certificate to that effect. That is because the issue of the final certificate is a procedure under the contract and the architect has the power only to do that which the contract empowers. Any settlement cannot be a settlement *under* the contract, but merely a settlement *of* the contract. The settlement should be separately recorded and signed by the parties as bringing the contract to an end. It is best done in the form of a deed to avoid any question that there is a lack of consideration. Proper legal advice is required.

Architects who take it upon themselves, or who are persuaded by clients, to issue final certificates for the amount of a settlement face the possibility of future challenges and the real risk that such certificates are invalid. Indeed, architects certifying in these circumstances are probably negligent in issuing certificates that they know to be wrong in the sense that they are not properly calculated in accordance with the contract.

129 If the contractor fails to provide the final account documents within the period specified in the contract after practical completion, what should the architect do?

SBC provides in clause 4.5.1 that the contractor must give the architect or quantity surveyor all the documents necessary for adjustment of the Contract Sum within 6 months after practical completion. The architect or quantity surveyor then has 3 months in which to ascertain any loss and/or expense and prepare a statement of adjustments to the Contract Sum. The Intermediate Building Contracts IC and ICD have clauses to similar effect.

Most contractors satisfy the requirement by submitting their own version of the final account, often earlier than practical completion. Many contractors, however, fail to provide all the documents necessary for substantiation. Delay in the issue of the final certificate can often be attributed to delays in the provision of this information. Of course, without substantiation, the quantity surveyor is hampered in completing the account. There have been many instances where the final certificate has been held up for literally years, because the information is not to hand. The Technology and Construction Court has considered this problem and given some useful guidance.[168] The JCT 80 form of contract was being considered but the principle holds good for these contracts also.

The court shone some much-needed light on the position when it pointed out that, if the contract had progressed properly, the information required by the quantity surveyor would have been obtained from the contractor during the progress of the Works. Strictly, the quantity surveyor should be keeping the status of the final account up to date throughout the contract period in accordance with any authorised variations. The effect of that would be that, by the time the certificate of practical completion was issued by the architect, the final account should be just about ready so far as the quantity surveyor was concerned. The purpose of clause 4.5.1 is to give the contractor:

> . . . a last opportunity to put its house in order and to ensure that the employer's representatives know of the full extent of the entitlement to which the contractor considers itself entitled and of the evidence to justify the amount of that entitlement.

The court said that if the contractor failed to take advantage of the opportunity, the architect and the quantity surveyor would have to do the best they could using whatever information the contractor has already provided together with their own knowledge of the project. The court made clear that the architect and the quantity surveyor cannot decline to act and, especially, the architect cannot refuse to issue a final certificate, because that would permit the contractor to control its issue and the contractor cannot be allowed to gain an advantage from its own breach.

Moreover, and perhaps more surprisingly, the court held that the provision to the contractor of a copy of the quantity surveyor's version of the final account was not necessary as a precursor to the issue of the final certificate. In the court's view, the final certificate itself

168 *Penwith District Council v V P Developments Ltd* (2005) 102 Con LR 117.

would be enough to allow the contractor to decide whether it was satisfied with the amount. The contractor was able to seek adjudication or arbitration if dissatisfied.

The ground rules are now clear. The quantity surveyor should keep the status of the final account up to date throughout the progress of the Works, seeking information from the contractor as required. After practical completion, the contractor has 6 months to submit anything further that may influence the final account. In any event, whether more information is submitted or not, the quantity surveyor should proceed with calculation of the final account after the 6 months expires, and the architect should issue the final certificate strictly in accordance with the contract. That provides in clause 4.15.1 that the final certificate shall be issued no later than 2 months after the last of the following three events: the end of the rectification period; the issue of the certificate of making good; or the date when the architect sends the copies of the final account to the contractor. The court made clear that the 2 months is a maximum period and, whenever the last event occurred, the final certificate could be issued the following day.

130 What does it mean to 'issue' a certificate?

This is a topic which causes considerable difficulty although it appears so simple. Many building contracts require certificates to be issued. It is referred to in SBC clause 1.9, IC and ICD clause 4.6.3 and MW and MWD clause 4.4. The date of issue of a certificate is very important. It triggers the period for payment and most contracts state that the final date for payment will be 14 days from the date of issue of the certificate. The employer is to send a notice stating the amount proposed to be paid no later than 5 days after the date of issue. Where the final certificate is concerned under SBC, IC and ICD, the date of issue marks the start of the 28-day period within which either party may commence adjudication, arbitration or other proceedings to challenge any or all of the content of the certificate. After the end of the period, the final certificate under those contracts becomes conclusive evidence (i.e. unchallengeable) in any future proceedings about a number of important points.

Although it has been the subject of some discussion, the date of issue of a certificate is established as being the date on which the certificate was sent out by the certifier. The fact that it is sent out, i.e. that it leaves the sender, is important. If the certificate was actually sent out from the architect on a later date than the date on the certificate, it probable that it would be the date of sending out (the date of issue) which would be relevant.[169] For example, it is not issued if it is merely signed by the architect and put in a drawer.[170] On the other hand, the issue of a certificate is not the same as the serving of a certificate on, or the receipt by, another party.[171] Issue and receipt may, and often are, achieved on different dates and it is significant that it is the date of issue, not receipt, which is important in JCT and other contracts. In *Token Construction Co Ltd v Charlton Estates Ltd*[172] the court said:

> [Counsel] stresses that clause 16 imposes no requirement of the document whereby 'the architect certifies in writing' his opinion must be served upon any particular person, and is thus unlike, for example, clause 21(a), which requires the architect to issue interim certificates to the 'contractor'. Nevertheless I have some difficulty in thinking that there would be sufficient compliance with clause 16 if the architect certified in writing and then locked the document away and told no one about it.

131 What can a contractor do if the architect fails to certify?

This can be a fairly tricky business. The architect is not a party to the contract, therefore, an important question is the extent to which an employer may be liable for the architect's failure to certify. An old case has decided that terms will be implied into a contract that an employer will not interfere with the architect's duties as certifier and that an employer will ensure that the architect carries out his duty as certifier.[173] Although this is an Australian case and not binding in the UK, it is thought that it properly represents the law here also. Moreover, a later English case held in regard to the architect:

169 *Cantrell & Another v Wright & Fuller Ltd* (2003) 91 Con LR 97.

170 *London Borough of Camden v Thomas McInerney & Sons Ltd* (1986) 9 Con LR 99.

171 *Glen v The Church Wardens & Overseers of the Parish of Fulham* (1884) 14 QBD 328.

172 (1973) 1 BLR 48.

173 *Perini Corporation v Commonwealth of Australia* (1969) 12 BLR 82.

> There may, however, be instances where the exercise of his professional duties is sufficiently linked to the conduct and attitude of the employer that he becomes the agent of the employers so as to make them liable for his default.[174]

Some doubt has been thrown on these fairly bold statements by a more recent case.[175] The court held that the architect, although employed by the employer, was given authority by both parties to the contract to form and express opinions and issue certificates as and when required by its terms. When so acting, the architect was not the agent of the employer, but:

> [the employer] was the party who could control [the architect] if he failed to do what the contract required. Since the contract is not workable unless the certifier does what is required of him, [the employer], as part of the ordinary implied obligation of co-operation, was under a duty to call [the architect] to book . . . if it knew that he was not acting in accordance with the contract. . . . the duty does not arise until the employer is aware of the need to remind the certifier of his obligations.

One thing the contractor must not do is to simply walk off the site and refuse to return until paid. That mistake was made by a contractor whose case was considered in *Lubenham Fidelities & Investment Co v South Pembrokeshire District Council and Wigley Fox Partnership*.[176] There, the Court of Appeal held that the employer was not in breach of contract in failing to pay sums in excess of what was shown on the certificates, because the employer's duty is merely to pay the amount certified. Moreover, the court held that if a certificate is erroneous, the proper remedy available to the contractor is to request the architect to make the appropriate adjustment in the next certificate or, if the architect declines, to take the dispute to arbitration.

It is important to distinguish the duty of the architect to certify from the duty of the employer to pay the amount certified. From the cases, it seems that action can only be taken against the employer if

174 *Rees & Kirby Ltd v Swansea City Council* (1983) 25 BLR 129.

175 *Penwith District Council v V P Developments Ltd* (2005) 102 Con LR 117.

176 (1986) 6 Con LR 85.

the employer knows that the architect has failed to certify. Therefore, it may be enough to found an action against the employer if, as soon as the certificate becomes overdue, the contractor writes to the employer notifying the fact and requesting that the employer take immediate steps to ensure that the architect complies with the contract. It would be necessary to allow a reasonable time for the employer to notify the architect and for the architect to certify, but in the context of a situation where a certificate is already late, a reasonable time could be 7 days. Indeed, it could be argued that, as the employer is party to the contract and ought, therefore, to know its payment terms, the employer would know that the architect had failed to certify at the same time as the contractor and should immediately take steps to rectify the situation. Nevertheless, a prudent contractor would first write to architect and employer requesting the certificate and giving a reasonable time for so doing. At the expiry of the time period, the contractor could refer the matter to adjudication for a speedy decision. Essentially the dispute would be that the architect had failed to issue the certificate and that the employer, in full knowledge and with express notice of such failure, had itself failed to ensure that certification took place. It should not be overlooked that SBC clause 13.7 and IC and ICD clause 4.8.6 provide that where there is a failure to issue an interim certificate on time or at all, the contractor's entitlement to interest is calculated from the day after what should have been the final date for payment.

132 Must the final account be agreed with the contractor before the final certificate is issued?

None of the JCT standard form contracts require that the final account is agreed with the contractor before the issue of the final certificate and neither does the ACA 3. The GC/Works/1 contract deals with the final account in clause 49 and although it makes provision for the contractor to agree or express any disagreement with the quantity surveyor's draft final account, clause 49(4) makes clear that where there is no agreement the account is to be determined in accordance with the contract. Therefore, the contractor's agreement is not necessary.

SBC clause 4.5 provides that not later than 6 months after practical completion, the contractor must provide all documents necessary for the adjustment of the Contract Sum. Not later than 3 months after receipt of such information, the architect or more usually the quantity surveyor, if so instructed by the architect, must calculate the final account and the architect must provide the contractor with a copy. Clause 4.15 deals with the timing and issue of the final certificate. The final account and final certificate provisions in IC and ICD are very similar and MW and MWD are similar, but much abbreviated. Nowhere does it suggest that the final account must be agreed by the contractor or that the final certificate cannot be issued until such agreement is obtained. Indeed, in SBC 1.9.2–1.9.4, the contract makes clear that disagreement with the final certificate, by either party, is only effective if made the subject of a referral to adjudication, arbitration or the courts as appropriate. The myth that the contractor has to agree the final account seems to have grown up due to the efforts of many quantity surveyors who, having had long discussions with the contractor about the final account, finally issue it to the contractor with a request that it indicates its agreement by signing and dating and then asks the employer to do the same. All this is of course quite unnecessary and can be misleading. Obviously, it is useful if the contractor indicates that it agrees with the final account sum, but the introduction of all the quasi-contractual agreement and signing procedure can overshadow the essential fact that the amount certified in the final certificate is a matter for the architect alone and quite irrespective of whether the contractor or the employer agrees with the figure. Sometimes, the employer effectively takes over the final account and meets the contractor to do a deal. Such deals are outside the contract machinery and where a deal is done in that way, there is no provision for the architect to issue a corresponding final certificate (see question 128). Moreover, it may well be negligent if the architect did so, because the final certificate represents the architect's professional opinion. Simply to reproduce in the certificate a figure agreed between the employer and the contractor can rarely if ever be the architect's professional opinion.

133 Is the final certificate ever conclusive about workmanship and materials?

It used to be thought that when the architect issued the final certificate it was equivalent to a statement from the architect that the whole of the Works had been carried out in accordance with the contract. That may well have been the position under early editions of the JCT contracts, but it is not the case now. There has always been an element of confusion about the final certificate. Some people believe that it is conclusive simply because it is the final certificate rather like grass is green because it is grass. It is not clear when this somewhat mystical approach to the final certificate took hold, but whenever it was, it was wrong. The first thing to understand is that the issue of the final certificate by the architect is simply a device adopted by the JCT and some other contracts to bring the contract to an orderly close. They could have adopted some other way. The final certificate under the MW and MWD contracts is not conclusive about anything at all, not even the final certified sum and after receiving payment of the balance, the contractor is at liberty to claim more money if it thinks it can substantiate the claim. However, it cannot do so under the contract mechanism, because the contract is at an end. The ACA 3 contract also has a final certificate and it also is not conclusive about anything. The conclusivity of the final certificate is peculiar to certain JCT contracts.

When a final certificate is said to be 'conclusive' what is meant is that it is conclusive evidence in regard to certain matters if neither party has referred such matters to adjudication, arbitration or other proceedings before the expiry of 28 days after its issue. A line has been drawn under those matters and they cannot be reopened. Therefore, if a final certificate is said to be conclusive that the amount of the final sum has been properly calculated, it will not prevent one of the parties from referring the method of calculation to adjudication or arbitration. But all that is necessary is for the other party to produce the final certificate to effectively have the claim dismissed. Certificates under SBC, IC and ICD are conclusive in respect of the following:

- *That if the contract bills or drawings or any architect's instruction or further issue of drawings states clearly that particular qualities of materials or goods or particular standards of workmanship are to be to the architect's approval, the particular quality or standards are to the architect's reasonable satisfaction:* It should be noted that the final sentence of this provision makes clear that the final certificate is not conclusive that any of those qualities or standards or any other materials, goods or workmanship comply with any other requirement of the contract. That means that, even if the architect has specified that something is to be to his or her satisfaction, it will not stop an employer seeking redress for work or materials which do not comply with the contract documents in other ways.

- *That all the provisions of the contract requiring adjustment of the Contract Sum have been complied with:* The final certificate is conclusive evidence that all necessary adjustments (adds, omits etc) to the Contract Sum have been properly carried out. Accidental inclusion or omission of work or materials, an obvious arithmetical error or fraud are the only matters that will allow the sum to be corrected.
- *That all extensions of time which are due under the extension of time clause have been given:* This prevents the contractor submitting new evidence about delays after the final certificate has been issued.
- *That reimbursement of loss and/or expense is in final settlement of all contractor's claims in respect of the matters identified under the loss and/or expense clause whether the claims are for breach of contract, duty of care, breach of statutory duty or otherwise:* This is a very broad clause intended principally, like the previous clause, to ensure that the final certificate really does signify the end of the road so far as finances were concerned. It should be noted that the conclusivity is effective only in respect of the loss and/or expense clause. It does not prevent the contractor from making claims in regard to breaches of contract which fall outside the scope of the matters.

Some confusion was generated when the effect of the issue of the final certificate as it applied to the architect's satisfaction with workmanship and materials was considered by the Court of Appeal.[177] Much to the concern of architects, the court interpreted the words 'which are to be to the satisfaction of the architect' very broadly and decided that all workmanship and materials were inherently matters for the architect's satisfaction and that the final certificate under JCT 80 was conclusive that the architect was satisfied with the quality and standards of *all* materials, goods and workmanship. The consequence was that the employer found it very difficult to take subsequent action against the contractor for latent defects. As a result, it became common for architects to avoid issuing a final certificate at all, leaving a small sum of money outstanding in the belief that the contractor would not bother to pursue it. It was hoped that the employer would not be prevented from seeking redress from the contractor if any defects appeared.

177 *Crown Estates Commissioners v John Mowlem & Co* (1994) 70 BLR 1 (CA).

Subsequently the JCT issued amendments to each of the affected forms of contract. The amendments were designed to remove the effect of the Court of Appeal decision by re-wording the sub-clauses relating to the architect's satisfaction. The position appears to have been restored to what everyone thought it was before the *Crown Estates* case. This was that the final certificate was conclusive about the architect's satisfaction only if the architect had specifically stated in the bills of quantities or specification that some item of goods, materials or workmanship was to be to his or her satisfaction or approval. The JCT 2005 series of contracts incorporate the amendment. The DB contract contains clause 1.8.1.1 which is to similar effect except that it refers to the final statement instead of the final certificate and to the employer rather than to the architect for obvious reasons.

Chapter 17

Sub-contracts

134 Must the architect approve the sub-contractor's 'shop drawings'?

It is not unusual for a contractor to submit a sub-contractor's or supplier's 'shop drawings' for approval before manufacture of the element concerned. Indeed, few sensible contractors would authorise proceeding with manufacture until the architect is satisfied with the details. Of course, in most cases the shop drawings are simply the sub-contractor's own translation of the architect's drawings and details into something that the sub-contractor believes is easier to understand in the context of the particular manufacturing process. In other words, the sub-contractor is using the information provided by the architect through the contractor to produce the shop drawings.

I once knew a very brave architect who would respond to the contractor with the following words: 'If the shop drawings are in accordance with the drawings I have provided, they are correct; if not, they are wrong.' This is equivalent to saying 'check them yourself'. It also requires a large degree of confidence on the part of the architect that the original drawings are completely accurate.

Few architects can say that their drawings are guaranteed to be 100 per cent correct. That is not to criticise architects; it is just a characteristic of the complex nature of the profession that discrepancies and other types of error do occur. Therefore, most architects will check shop drawings just to be sure that their own drawings are correct. The problem is that, in checking whether the shop drawings accurately represent their drawings, architects inevitably check things that have been introduced by sub-contractors. Sometimes, sub-contractors will actually change architects' details to make them suit the particular sub-contract element. Such changes can easily be missed if the architect gives the drawings only a cursory inspection. Architects should either check shop drawings thoroughly or not at all. Even if the architect has no contractual responsibility for checking such drawings, responsibility may be assumed if the architect nonetheless does check them.

In most cases, the architect will want to be satisfied that the shop drawings are accurate and, therefore, will check them. Whether the architect has an obligation to approve the drawings will depend upon the terms of the contract. Such an obligation will usually be found, if at all, in the preliminaries section of the bills of quantities or specification. Ideally, the architect should make sure, before the documents are sent out for tender, that there is no requirement for the approval of the architect. The absence of the requirement for approval will not prevent the contractor from sending the drawings for approval, but it will enable the architect to point out that there is no contractual requirement for the architect's approval. Moreover, the architect should inform the contractor that it is the contractor's task to check and co-ordinate sub-contractors' drawings.

Obviously, if the sub-contractor is being asked to carry out part of the design, the position is rather different. The architect, who is usually the design leader, will have a duty to co-ordinate the sub-contractor's design with the rest. Therefore, the architect will have a corresponding duty to check the drawings to ensure this co-ordination.

The position is, therefore, clear. The architect will rarely have any obligation to approve a sub-contractor's shop drawing unless either the sub-contractor has a design obligation or the contract documents expressly require the architect to approve such drawings. When dealing with the sub-contractor's design, it is safest if the architect avoids using the word 'approve' and instead simply states that he or she has no comment to make. Use of the word 'approve' has been discussed elsewhere. It does not usually remove any responsibility from the contractor.

135 MW: If the contractor is in financial trouble, can the employer pay the sub-contractors directly?

Under JCT 98, there used to be provision for the employer to pay nominated sub-contractors directly in certain circumstances. There are no such provisions in SBC; indeed, there are no nominated sub-contractors in SBC. Even under JCT 98 terms, the direct payment provisions were hedged around by substantial conditions.

Under current traditional JCT contracts, there are no circumstances where the employer should pay sub-contractors directly.

It is important to understand that the employer is in contract with the contractor and the contractor is in contract with the sub-contractors. There is no contractual relationship between the sub-contractors and the employer unless some kind of direct warranty has been employed, because clauses in the main and in the sub-contracts prevent the Contracts (Rights of Third Parties) Act 1999 having any effect. Therefore, the position is that the contractor has undertaken to the employer, for payment, to carry out certain Works. Part of these Works has been sub-let to sub-contractors. That is to say, the sub-contractors have each undertaken to the contractor to carry out their parts of the main contract Works in return for payment from the contractor.

When a sub-contractor carries out work for the contractor, it is part of the main contract Works and the contractor is entitled to payment for it from the employer. If the contractor does not pay the sub-contractor, the sub-contractor's redress is against the contractor. The sub-contractor has no valid claim directly against the employer. It is unfortunate for the sub-contractor (indeed for all concerned) if the contractor gets into financial difficulties or even goes into liquidation. That is the kind of thing that the sub-contractor, like any other business, must try to guard against. Some employers believe that they are entitled to pay the sub-contractor directly and then deduct the money paid from the contractor. That is wrong.

The employer who pays directly will be in breach of the insolvency rules by making the sub-contractor into a preferential creditor. Even if that is not an issue, in the case of a contractor who simply will not pay sub-contractors, the employer will undoubtedly be called upon to pay the contractor in any event for the work carried out as part of the main contract by the sub-contractor. The employer will have no defence. It is not an argument for the employer to say: 'I will not pay you because you have not paid your sub-contractors.' The employer's duty to pay the contractor under the main contract is not dependent on whether the contractor has paid the sub-contractors. Indeed, the contractor's relationship with sub-contractors is no business of the employer's except to the extent that the main contract requires the contractor to include certain provisions in the sub-contract (for example, SBC clause 3.9.2).

On a purely practical level, there is no way in which the employer can be sure that the sub-contractor has not been paid unless the sub-contractor takes legal action against the contractor. In that case, the sub-contractor will recover whatever the adjudicator, arbitrator or judge believes is appropriate.

136 If the architect instructs the contractor to accept a specific sub-contractor's quotation, is the employer liable if the contractor fails to pay?

The question of whether a contractor can refuse to accept such a quotation has been dealt with elsewhere. Assuming that a contractor has accepted the quotation, the question really is whether the employer has any liability to the sub-contractor after that. It often happens that a contractor becomes insolvent while owing money to sub-contractors. Does it make a difference that the contractor was instructed to accept this particular sub-contractor? In principle, it does not make a difference. So far as the employer is concerned, this is a sub-contract freely entered into between contractor and sub-contractor by which, in return for work, the contractor agrees to pay the sub-contractor a certain amount of money. If, for any reason, the contractor fails to pay, that is a matter between the contractor and the sub-contractor. The employer is not a party to the sub-contract and has no liability to either party under it although of course the employer has certain duties to the contractor under the main contract. This question was probably prompted by memories of the nominated sub-contract provisions in JCT 98. Under those provisions, if the contractor failed to pay a nominated sub-contractor money which the contractor had been directed, as part of the certificate, to pay to the nominated sub-contractor, the architect could certify the failure to the employer who then had to pay the sub-contractor directly the amount owed by the contractor. The employer was entitled to deduct the amount from any further money due to the contractor. Even these provisions were modified if the contractor became insolvent. There are no nominated sub-contract provisions in the 2005 suite of JCT contracts and the existing naming provisions do not allow the employer to pay sub-contractors directly if the contractor fails.

137 If the contractor engages a sub-contractor without the architect's consent, can the

contractor avoid having to pay the sub-contractor for work done?

SBC clause 3.7.1 states that the contractor must not sub-contract any part of the Works without the architect's consent which must not be unreasonably delayed or withheld. The provision is important, because when the employer, with the advice of the architect or the quantity surveyor, selected contractors for the tender list and when the employer accepted the price of one of them, it was on the basis, at least partly, of the reputation of those contractors. If a contractor was allowed to sub-contract as it pleased and to choose sub-contractors indiscriminately, perhaps simply on the basis of the lowest price, the point of selecting a contractor with a reputation for good work would be defeated. If a contractor sub-contracts without consent, it is a ground for the issue of a default notice under clause 8.4.1.4 prior to termination by the employer.

In practice, a contractor will often sub-contract without consent and then request consent later. Many architects feel pressured into giving the consent simply because the consequences of withholding consent, even if justified, might be to cause a severe delay to the contract. Although the delay would be the contractor's fault, and for which the employer could recover liquidated damages, the actual period of delay could never be recovered. Therefore, the instances when an architect will refuse consent to a sub-contracting are likely to be rare. Nevertheless, it does occur and when it does, the contractor may be in the position of having instructed the sub-contractor to carry out part of the work, which may have to be redone with an authorised sub-contractor or if sub-contracting is not consented to at all, by the contractor itself. Even worse, the contractor may have entered into a sub-contract with the sub-contractor for the whole of that particular item of work, for example heating, and face liability for repudiation of the sub-contract.

Does the refusal of consent by the architect, necessitating the use of a different sub-contractor by the contractor or the carrying out of that work by the contractor itself, somehow invalidate the sub-contract so as to relieve the contractor from any liability under it. Sadly for the contractor, the answer to that question is 'No'. The obligation to obtain consent prior to sub-contracting is a term of the main contract between the contractor and the employer. If the contractor enters into a sub-contract with the sub-contractor without consent, that sub-contract is perfectly valid and legally binding on both sub-contractor and contractor. The terms of the main contract are of no concern to the sub-contractor unless the sub-contractor has received adequate notice of them. Even if such notice has been given, the sub-contractor has no duty to check with the contractor that consent has been received from the architect. The sub-contractor's position would be the same as that of the party to any other contract where repudiation had taken place. It would be impossible for the sub-contractor to continue and it would have little choice but to accept the repudiation and claim damages.

138 The contractor has gone into liquidation and the heating sub-contractor says it is going to remove all the loose piping stored on site and take away the radiators fixed in the building. Can it do that?

This is a complex topic. Very often, when a contractor goes into liquidation, sub-contractors and suppliers are left unpaid. SBC attempts to deal with this situation in clauses 2.24 and 3.9. Clause 2.24 states that site materials must not be removed from the Works without the architect's consent and that if the value has been included in an interim certificate and the amount has been paid by the employer to the contractor, the materials become the property of the employer. Of course, this clause is binding only between the employer and the contractor. It cannot bind a third party such as a sub-contractor or supplier. The contract tries to overcome that problem in clause 3.9 where it states that 'when it is considered appropriate', whatever that may mean, the contractor shall enter into the relevant version of the JCT standard sub-contract. The clause then proceeds to state what the sub-contract must include. Among the terms included are terms which reflect clause 2.24 stating that site materials must not be removed from the Works without the architect's consent and that if the value has been included in an interim certificate and the amount has been paid by the employer to the contractor, the materials become the property of the employer. While the contractor remains solvent and provided that the contractor has complied with clause 3.9 of SBC, removal of materials from the site by a sub-contractor would be a serious breach of the sub-contract for which the contractor would be liable to the employer through the main contract.

The difficulty arises when the contractor becomes insolvent, because there is usually no point in taking legal action for breach of contract against an insolvent party. If the contractor has failed to ensure that the requisite sub-contract has been executed, it is a breach of contract with no remedy. Even if the insolvent contractor has ensured the execution of the sub-contract, the contractual exclusion of rights under the Contracts (Rights of Thirds Parties) Act 1999 (clause 1.6 of SBC and similar clauses in other JCT contracts) means that it confers no benefit on the employer and the sub-contractor can ignore it with impunity. Many suppliers and sub-contractors will have a retention of title clause in their contracts with the contractor. This clause states that the sub-contractor retains ownership of the materials until it receives payment. Therefore, it owns its materials on site if not paid for. If it has entered into a standard JCT sub-contract with the contractor, the retention of title clause will not usually apply. For the sub-contractor to enter the site and remove loose materials for which it has not been paid would amount to trespass. However, if the trespass has been carried out without causing any damage to the Works or the site, it is difficult to see what the employer can do about it. The employer's best short-term solution is to make the site and all materials secure as soon as the insolvency is confirmed. However, despite SBC clause 2.24, the sub-contractor may still have a valid claim to the loose materials.

Fixed materials such as radiators are an entirely different matter. In principle, once the material has been fixed to the building, it becomes part of the building and, therefore ownership passes to the employer.[178] Radiators securely fixed to the building are no longer the property of either sub-contractor or contractor. They belong to the employer and if the sub-contractor removes them it is theft. Where such removal is threatened, it is important that the employer or the employer's solicitor writes immediately to the sub-contractor informing it that it is threatening to commit a criminal offence.

139 Can there be liquidated damages in a sub-contract?

The straight answer to that question is that there can be provision for liquidated damages in a sub-contract, but it is not usually a very good idea. There is a great deal of case law concerning liquidated damages stretching back many years. For example, a leading case is *Dunlop*

178 *Reynolds v Ashby* [1904] AC 466.

Pneumatic Tyre Co Ltd v New Garage & Motor Co Ltd.[179] These cases deal with attempts to impose liquidated damages in all kinds of circumstances. What is clear is that the purpose of liquidated damages is to avoid the need for parties to become involved in expensive litigation in order to establish a breach of contract, prove the damage resulting from that breach and then quantify the cost of rectifying the damage. Instead, the parties agree on a sum of money which is to be deducted on the occurrence of an event. Then all that need be done is to identify that the event has occurred and to allocate the sum of money. In building contracts, liquidated damages are almost invariably connected to a failure by the contractor to complete the Works by the completion date in the contract. Liquidated damages are usually stated as £x for every day or week that the Works are delayed. Before the liquidated damages can lawfully be deducted, it must be shown that the contractor was not entitled to any extension of the contract period. The periods of time between contract completion date and actual (practical) completion are matters of fact.

A key factor in dealing with liquidated damages is that the sum of money must be a genuine pre-estimate of the loss likely to be suffered by the employer if delays occur. A genuine pre-estimate does not mean that the sum has to be calculated with great precision. Sometimes precision can be used and, where it can, it should. However, in most cases, the employer may only have a vague idea, at the execution of the contract, of the sum likely to be lost. In general terms, provided a genuine effort has been made no matter how guessed the outcome, the damages will be enforced by the courts. In calculating the sum, the employer has to think of only one thing – the consequences if the contractor does not finish on time. Failure to finish will be instantly clear.

Contrast that situation with a sub-contractor. A sub-contractor is one of several, perhaps many, on the site. A sub-contractor has its own timescale in which to work, in and amongst other sub-contractors on the site. It is often necessary for a sub-contractor to go away and then return to site when needed. Therefore, when looking at the programme encompassing all the work to be done on a project, all the sub-contractors are closely linked and due to complete their sections of the Works at different times. If a sub-contractor is delayed due to its own fault, the delay will affect some, but probably not all

179 [1915] All ER Rep 739.

of the other sub-contractors. A sub-contractor in delay can do little to recover lost time, because the delay will affect others and even if the delaying sub-contractor returns to site later and finishes that piece of work in half the time, it is unlikely to have a major or any effect on the Works as a whole. The chances are that the main contractor will also cause some delays which will obviously affect the sub-contractors. It is impossible for the main contractor to calculate a suitable sum for each sub-contractor to represent a genuine pre-estimate of the loss likely to be sustained by the main contractor if any or all of these sub-contractors fail, through their own fault, to meet their individual completion dates. Even if it was known that just one of the sub-contractors would fail to meet its completion date, the calculation of resulting loss to the main contractor would be difficult to do after the event, but certainly not possible before. Therefore, the sub-contractor should be able to successfully challenge the whole basis and amount of liquidated damages. Where several sub-contractors are in delay, the problem is magnified. Main contractors have been known to insert, as the liquidated damages for each sub-contractor, the amount of liquidated damages which the main contractor is liable to pay under the main contract. It takes but a moment's thought to see that such an approach is completely untenable.

It is no coincidence that all the standard form sub-contracts which have been produced provide for unliquidated damages only. In other words, they all acknowledge that the main contractor will have to prove that a sub-contractor was late through its fault and the amount of damages suffered thereby.

140 Is the sub-contractor obliged to work in accordance with the actual progress of the main contractor's Works?

Prior to 2005, standard sub-contracts tended to require the contractor to carry out and complete the sub-contract works in accordance with the programme details in the sub-contract and reasonably in accordance with the progress of the Works. Many contractors assumed that, having entered into a sub-contract with a sub-contractor, the sub-contractor was obliged to work when and where on the Works the contractor directed. Contractors expected to be able to telephone the sub-contractor and to secure its presence on site virtually immediately and that it would move off the site with equal speed as directed. When tested in the courts, that approach was held to be wrong.[180] The courts decided that the sub-contractor could plan and carry out the work as it wished if the sub-contract said nothing to the contrary, but the sub-contractor must finish by the time stated in the sub-contract. The sub-contractor was not required to perform the work in any particular order or at any particular rate of progress or to finish any part of the sub-contract works by any particular date so as to enable the contractor to proceed with other parts. But the sub-contractor must not unreasonably interfere with the carrying out of other work which it was convenient to carry out at the same time. The JCT 2005 suite of sub-contracts have wording which is slightly changed from previous editions, but it is thought that the change in wording does nothing to increase the sub-contractor's obligations so far as progress is concerned, but rather confirms the position as set out by the court. The situation is not entirely one-sided. A sub-contractor has argued that a term should be implied into sub-contracts that the main contractor would make sufficient work available to enable the sub-contractors to maintain reasonable progress and to execute the work in an efficient and economic manner. That argument has been rejected by the Court of Appeal, which held that the general law had no rule implying such a term into sub-contracts.[181] Everything depends on the precise terms of the sub-contract being used. Often such sub-contracts are not in standard form but rather sub-contracts specially written for the contractor.

180 *Pigott Foundations v Shepherd Construction* (1994) 67 BLR 48.
181 *Martin Grant & Co Ltd v Sir Lindsay Parkinson & Co Ltd* (1984) 3 Con LR 12.

Chapter 18

Extensions of time: adjustment of the completion date

141 Can the architect ignore delays if the contractor has failed to give proper notice?

This is a common question which depends on the terms of the contract. At one time it was considered that if the employer committed any act of prevention, the contractor was entitled to an extension of time irrespective of any notice provisions and if no extension was given, time became at large and the contractor's obligation was simply to complete the Works within a reasonable time.

It has since been recognised that such an approach would allow a contractor to put time at large at will by the simple expedient of ignoring notice provisions which would have triggered an extension of time. It has been held that if a contractor ignores a notice provision which is a condition precedent, there will be no entitlement to extension of time, even though there would otherwise be a clear basis on the grounds of the employer's acts of prevention.[182] It has been said that in order for a requirement for notice to be a condition precedent, several serious conditions must be satisfied. For example, the time for service of the notice must be stated and also must make clear that if there is a failure to give the notice, a specific right will be lost. That position has been considered by the court and it was rejected expressly in regard to extension of time where there was a clause which had as a proviso the words: 'the sub-contractor shall have given within a reasonable period written notice to the contractor of the circumstances giving rise to the delay'.[183] The court held that:

182 *Multiplex Constructions (UK) Ltd v Honeywell Control Systems Ltd* (2007) 111 Con LR 78.

183 *Steria Ltd v Sigma Wireless Communications Ltd* (2007) 118 Con LR 177.

> if there is a genuine ambiguity as to whether or not notification is a condition precedent, then the notification should not be construed as being a condition precedent, since the provision operates for the benefit of only one party i.e. the employer, and operates to deprive the other party (the contractor) of rights which he would otherwise enjoy under the contract.
>
> . . . in my judgment the phrase . . . is clear in its meaning. . . . In my opinion the real issue which is raised on the wording of this clause is whether those clear words by themselves suffice, or whether the clause also needs to include some express statement to the effect that unless written notice is given within a reasonable time the sub-contractor will not be entitled to an extension of time.
>
> In my judgment a further express statement of that kind is not necessary. I consider that a notification requirement may, and in this case does, operate as a condition precedent even though it does not contain an express warning as to the consequence of non-compliance.

In *City Inn Ltd v Shepherd Construction Ltd*,[184] the contractor was required to carry out various actions and make specific submissions before being entitled to an extension of time in regard to certain architect's instructions. The court had no hesitation in holding that failure to comply with the requirements would prevent the contractor from getting any extension of time in respect of the particular instructions. It is quite common to find a requirement for notice in standard form contracts, particularly in regard to extensions of time and applications for loss and/or expense. These recent decisions suggest that the courts will not ignore a contractor's failure to give notices prescribed by the contract and it seems likely that such clauses will be interpreted as conditions precedent. If that is correct, contractors will have to give timely and effective notices before the architect need consider them.

142 Is a note in the minutes of a site meeting sufficient notice of delay from the contractor?

From time to time, questions arise about the status of site meeting minutes. A favourite question is whether they constitute a written

184 [2008] BLR 269.

instruction of the architect as required by the JCT and other standard contracts. The answer is probably that they do, provided that the minutes are drafted by the architect. A more bizarre question is whether minutes are written applications by the contractor in respect of direct loss and/or expense. The answer to that is clearly that they are not, even if they record that the contractor asked for loss and/or expense at the meeting, unless the contractor actually drafted the minutes.

A related question is whether a note in site meeting minutes could be a notice of delay to comply with clause 2.27.1 of SBC. The clause requires the contractor to give written notice forthwith whenever it becomes reasonably apparent that the progress of the Works is being or indeed likely to be delayed. This question was given short shrift by the court in a recent decision:[185]

> I also consider that the written notice must emanate from [the sub-contractor]. Thus for example an entry in minute of a meeting prepared by [the project manager] which recorded that there had been a delay . . . would not in my judgment by itself amount to a valid notice under cl.6.1. The essence of the notification requirement in my judgment is that [the main contractor] must know that [the sub-contractor] is contending that relevant circumstances had occurred and that they have led to a delay in the sub-contract works.

In that case, an amended MF/1 form of contract was used between main contractor and sub-contractor, but the principle of a written notice was similar to what is required under JCT contracts. It clearly cannot be the contractor's written notice if it is in a document written by someone other than the contractor. However, there seems no reason to doubt that if a progress report is presented in writing to the architect at a site meeting, that report would suffice as a written notice provided that it contained the relevant information required under the contract. Any reference to it in the site meeting minutes would be no more than a reference. The notice would be the report itself.

143 Is time of the essence in building contracts?

Time is 'of the essence' when any breach of stipulations about time in a contract can be treated as repudiatory so as to entitle the other

185 *Steria Ltd v Sigma Wireless Communications Ltd* (2007) 118 Con LR 177.

party to terminate performance of its obligations and claim damages. There are probably only three instances where time will be of the essence:

- if it specifically states in the contract that time is of the essence;
- if time being of the essence is a necessary implication;
- if one party is in delay to an unreasonable extent, time may be made of the essence if the other party serves a notice on the party in breach setting a new and reasonable date for completion.[186]

The term must be so fundamental that a failure to comply would make the contract almost worthless so far as the other party was concerned. An example might be if a shop orders goods for sale to meet a particular demand and that demand is known to the supplier, failure to meet the required delivery date might, depending on all the circumstances, amount to a breach of an essential term.

Time will not normally be of the essence in building contracts unless expressly stated to be so. The reason is that the contract makes express provision for the situation if the contract period is exceeded in the form of an extension of time clause and liquidated damages.[187] Therefore, it would be a nonsense to make time of the essence: one term would be to the effect that failure to complete by the due date would entitle the other party to accept the failure as repudiation, while the other term would allow the contractor to an extension of time for appropriate reasons. Time was made of the essence in *Peak Construction (Liverpool) Ltd v McKinney Foundations Ltd*[188] and apparently gave the employer the right to terminate the contract at the end of the period as extended by the architect. In the case of most building contracts, the provisions for termination (e.g. for failure to proceed regularly and diligently) adequately cover the situation.

144 What, in practice, are 'concurrent delays'?

Concurrent delays are the subject of much debate and, it must be said, a great deal of misunderstanding. A simple approach is to say that it is when two or more things happen at the same time and delay

186 *Shawton Engineering Ltd v DGP International Ltd* [2005] EWCA Civ 1359.
187 *Babacomp Ltd v Rightside* [1974] 1 All ER 142.
188 (1970) 1 BLR 111 (CA).

the completion date of a particular contract. However, that is not the whole story: there are two kinds of concurrent delays. There are delays which occur at the same time to two different activities and there are the delays which occur at the same time to the same activity. It is the latter which causes the problems.

If delays act on different activities, it is easy to deal with the situation by inputting the delays, one at a time, into the contractor's programme and noting the effect. This can be done using computer software. It is only the delays which are grounds for extension of time which must be entered. Contractor's culpable delays are ignored. One court commented that:

> . . . it is agreed that if there are two concurrent causes of delay, one of which is a Relevant Event, and the other is not, then the contractor is entitled to an extension of time for the period of delay caused by the Relevant Event notwithstanding the concurrent effect of the other event.[189]

This must be a correct view where the two concurrent delays operate on different activities and, as noted earlier, computer analysis easily deals with this situation by ignoring all delays which are not relevant events.[190]

True concurrency is when the causes of delay operate at the same time on the same activity. To take a simple example, a contractor may have difficulty in obtaining labour to lay paving, but the architect may also be delayed in providing the necessary drawings showing the layout of the paving. The courts have not adopted an entirely consistent approach. Indeed, the guidance one might expect from the courts has not been very helpful. Sometimes, what is known as the 'dominant cause' approach has been advocated. One court has asked '. . . what was the effective and predominant cause of the accident that happened, whatever the nature of the accident may be'.[191] It has been said that the test is that of the ordinary bystander. Yet again, a court has suggested that the architect might simply apportion the responsibility for delay between two causes. As to how that is to be done, the court concluded that the basis must be fair and

189 *Henry Boot (Construction) Ltd v Malmaison Hotel (Manchester) Ltd* (1999) 70 Con LR 32.

190 *Steria Ltd v Sigma Wireless Communications Ltd* (2007) 118 Con LR 177.

191 *Yorkshire Dale Steamship v Minister of War Transport* [1942] 2 All ER 6.

reasonable.[192] A sensible view of concurrency was given in a later case:

> However, it is, I think, necessary to be clear what one means by events operating concurrently. It does not mean, in my judgment, a situation in which, work already being delayed, let it be supposed, because the contractor has had difficulty in obtaining sufficient labour, an event occurs which is a Relevant Event and which, had the contractor not been delayed, would have caused him to be delayed, but which in fact, by reason of the existing delay, made no difference. In such a situation although there is a Relevant Event, '*the completion of the Works is [not] likely to be delayed thereby beyond the Completion Date.*' The Relevant Event simply has no effect upon the completion date.[193]

True concurrency is fortunately rare. In most instances, where delays occur to the same activity, they are consecutive.

145 How does time become 'at large'?

In the absence of an express term in a contract fixing the date for completion, the contractor's obligation is to complete within a reasonable time. (The task of determining what a reasonable time is on any particular occasion is a question of fact depending on all the terms of the contract and the surrounding circumstances.) That is often referred to as time being at large. Time is at large when there is no date for the completion of the contract or if the dated fixed has become inoperable for some reason.

All the standard forms of building contract provide for a date for completion and also make provision for fixing a new date for completion where delay has occurred due to the actions or inactions of the employer or someone for whom the employer is responsible, for example, the architect. It is relatively rare for time to become at large and, possibly for this reason, many adjudicators appear reluctant to come to that conclusion even where all the evidence points to it. The usual reason why time may become at large is because the architect

192 *City Inn Ltd v Shepherd Construction Ltd* [2008] BLR 269.

193 *Royal Brompton Hospital NHS Trust v Hammond and Others (No. 7)* (2001) 76 Con LR 148.

fails properly to give an extension of time under the contract provisions.

It will be unusual for time to become at large if the architect properly operates the contract. One of the consequences of time becoming at large is that liquidated damages can no longer be deducted because there is no date from which such damages can be calculated.[194]

The position can become complex if the architect awards extensions of time after completion of the Works. Whether or not the architect is entitled to do so depends on the precise wording of the contract, construed in the light of decided cases. SBC, IC and ICD expressly give the architect power to extend time and fix a new date for completion after the contract completion date has passed.

146 Under SBC, if the architect gives an instruction after the date the contractor should have finished, is the contractor entitled to an extension of time, and if so how much?

It is often argued, with some merit, that the architect should issue all instructions before the completion date in the contract. But what is the situation if the completion date has passed and the contractor has been trying to finish the Works for several weeks when an instruction for additional work is issued? Clearly, the instruction is going to cause a delay and the contractor is entitled to an extension of time as a result. The question is whether the contractor is entitled to an extension of time from the completion date to the date on which the contractor actually finishes the additional work or it is entitled merely to an extension of time added to the last completion date of a length to represent the time taken to comply with the instruction.

The contractor will usually be looking for the former, but the architect will be anxious to give the latter. The contractor's argument makes sense. It amounts to this: the architect could have issued the instruction at any time up to the date of completion, but chose to issue it after the completion date was past. There is no excuse for that, because where SBC is used, all the work to be done should be known when the Works commence, therefore the architect cannot say that he or she could not issue the instruction until the contractor had reached a particular stage.

194 *Wells v Army & Navy Co-operative Society Ltd* (1902) 86 LT 764.

The architect's argument is simply that instructions may be issued at any time up to practical completion, therefore although an extension of time is due, it can deal only with the actual period of delay. The contractor is effectively saying, 'How can you give me an instruction to carry out work today knowing that it will take until next week to do, but at the same time you are giving me an extension of time that says I should have finished several weeks ago?'

The question was considered by the court in *Balfour Beatty Ltd v Chestermount Properties Ltd*[195] in connection with the JCT 80 form of contract. The court decided that the contractor was entitled only to the net amount of the delay added on to the completion date. The court's approach appeared to be that it was not the relative dates which were important, but rather the time periods of the delay. This did not seem to be a result of a strict reading of the contract but rather the court's view of what was a reasonable solution to the problem. Essentially, the court said that in such circumstances, the architect should estimate the length of time the contractor needed to complete the work in the instruction and then add it on to the last date for completion. This had the effect of removing the contractor's period of culpable delay from the extension period. It can be seen that this is a just solution, but whether it is what the contract actually said is open to doubt. The decision was based on the JCT 1980 and it is applicable to SBC, IC and ICD.

147 If, under SBC, the architect does not receive all the delay information required until a week before the date for completion, must the extension of time still be given before the completion date?

This question originates from the fact that clause 2.28.2 states that the architect must notify the contractor of the decision in writing 'as soon as is reasonably practicable' but it must be within 12 weeks of receipt of the required particulars. The clause goes on to state that, if the period between receipt and the completion date is less than 12 weeks, the architect must endeavour to notify the contractor before the completion date. This wording, and similar wording in the predecessor contract JCT 98, is a source of concern to many architects.

195 (1993) 62 BLR 1.

The current wording stipulates that the architect must 'endeavour' to notify before the completion date. That is new wording introduced with SBC and requires the architect to make an earnest or determined attempt. The previous wording let the notification depend on whether it was 'reasonably practicable' to do so.

The key is the date of receipt of the required particulars. The particulars are detailed in clause 2.27.2 and are said to include the expected effects of the relevant event and an estimate of any expected delay. It is only when the architect receives this information that the time period for making a decision begins. Therefore, if the architect does not receive the particulars until a week before the completion date, the architect is still expected to 'endeavour' to reach a decision and notify the contractor before the completion date; but realistically it is probably unlikely.

Contractors sometimes delay providing the information until there are only days to go before the completion date in the hope of being able to argue that the architect has failed to act and, therefore, time is at large. Such arguments have little chance of success. Indeed, if the architect cannot reach a decision before the completion date, despite endeavouring to do so, he should do so as soon as possible. Alternatively, it may be considered that the matter is then part of the decision to be made by the architect under clause 2.29.5. This clause states that after the completion date the architect *may* and not later than 12 weeks after the date of practical completion the architect *shall* (must) carry out a review of all extensions and make what is effectively a final decision. Therefore, if the contractor delays in providing the required particulars, it runs the risk that a decision will not be made in respect of those delays until some weeks after practical completion.

There is nothing that obliges the contractor to provide additional information requested by the architect but, obviously, if the architect reasonably asks for information to assist in making the decision the contractor would be foolish to withhold it. For example, the architect may say that it is not possible to understand the effects unless the contractor provides a critical path network showing the effect of the delay on the completion date. That does not appear to be an unreasonable request. On the contrary, it seems to be a request in the interests of all parties.

148 Is there an easy way to decide an extension of time?

The simple answer to that is that there is no easy way, but that some ways are easier than others. The ways in which delays can operate on a contract are many and varied and trying to assess the effect of a dozen or more delays acting on different activities, some acting on the same activity, is quite difficult. For many years, architects would try to solve the conundrum by comparing actual against programmed progress on a bar chart, making what allowance seemed appropriate for contractor's own delays and finally making a stab at it.

Some architects, with total disregard for the contract provisions, approached the problem from an entirely new angle and simply considered whether they felt that, taken as a whole, the contractor's progress was reasonable or not and decided upon the extension of time accordingly. Needless to say, this kind of exercise usually resulted in contractors being given more extension of time than their strict entitlement and occasionally to rather less than their entitlement.

In recent years, more reliance has been placed on the computer and its ability to perform a great many complex calculations almost instantaneously. Although the initial effort required is fairly substantial, the advantage of using a computer to assist in calculating extensions of time is that a degree of logic can be brought to the process and the analysis can be carried out on a reasonable basis. The courts have shown themselves ready to accept such analysis provided that it is properly carried out and it is certainly better than when the architect reaches a decision based on an impressionistic assessment.[196] A computer program based on a critical path network or precedence diagram is used and there are a number of such programs available. The programs enable activities to be linked in a logical fashion so that the effect of a delay on one activity can be tracked through the building programme to the completion of the works. The programs allow activities to be delayed or started early and resources can be added. Architects and project managers should routinely use computerised programmes as a matter of course to monitor progress and assist in analysing delays. Contractors should submit detailed programmes on disk as well as in hard copy.

196 *John Barker Construction Ltd v London Portman Hotels Ltd* (1996) 50 Con LR 43.

Programmes can be prepared to show the progress as-built compared to intended progress and known employer-generated delays can be taken out to examine the likely situation had those delays not occurred. The reverse operation can be tried. The logic links will determine the effect of delays, and it is important to ensure that they are shown correctly.

In *Balfour Beatty Construction Ltd v The Mayor and Burgesses of the London Borough of Lambeth*[197] reference was made to the use of programmes for estimating extensions of time. As part of its submission in adjudication, Balfour Beatty referred to the 'most widely recognised and used' delay analysis methods:

(I) Time Impact Analysis (or 'time slice' or 'snapshot' analysis). This method is used to map out the impacts of particular delays at the point in time at which they occur permitting the discrete effects of individual events to be determined.
(II) Window analysis. For this method the programme is divided into consecutive time 'windows' where the delay occurring in each window is analysed and attributed to the events occurring in that window.
(III) Collapsed as-built. This method is used so as to permit the effect of events to be 'subtracted' from the as-built programme to determine what would have occurred but for those events.
(IV) Impacted plan where the original programme is taken as the basis of the de lay calculation, and delay faults are added into the programme to determine when the work should have finished as a result of those delays.
(V) Global assessment. This is not a proper or acceptable method to analyse delay.

In its most basic form, when dealing with JCT contracts, the architect should list all the delays notified by the contractor and consider each one to determine whether it is a relevant event and, if so, the amount by which the activity in question has been delayed. After putting the contractor's planned critical path programme on the computer, each delay caused by a relevant event should be inputted into the programme and the results observed. The computer-calculated new end date will be the date most favourable to the

197 [2002] BLR 288.

contractor, because in reality the contractor would probably carry out some reprogramming after each delay occurred, thus reducing the overall effect of subsequent delays. But this will provide a relatively quick answer and one that will demonstrate how the delays have affected progress. The hard work comes in putting the programme data into the computer and the architect's judgment comes into play in determining the amount of delay to each activity as a starting point.

149 If there is a clause in the contract which says that the employer will remain in residence during alterations to a house, but the employer in fact moves out, should the improved working conditions count as a 'discount' against any extension of time which might be due?

Most contracts provide for the contractor to have exclusive possession of the site and such a term would usually be implied in any event. Therefore, if the employer is going to remain in occupation of some or all of the property, a term must be included in the contract to that effect so that the contractor can make a suitable allowance for any likely delays or inefficiencies arising. The contractor might decide to make such allowance by requiring a longer contract period than would be the case if the employer was not in residence, or the contractor may add something to its price or both.

If the employer subsequently decides to leave the premises, that is entirely a matter for the employer. The contractor cannot be required to adjust its price or contract period as a result. Indeed, it is difficult to see how the extra time and/or money allocated by the contractor to working while the employer was in residence can be determined unless the contractor volunteered the information and it is under no obligation to do so. The contractor has agreed to contract on the basis that the employer would stay in residence, therefore, the employer has the option. There has certainly been a change in circumstances, but no change in the work to be done.

Although the improved working conditions cannot be set against any extension of time otherwise due, it is likely that the architect would be entitled to take into account the contractor's current working conditions when calculating whether the contractor has been delayed. In other words, the architect would not be obliged to artificially assume that the contractor's work methods were already impeded by the presence of the employer when a delay occurred: the architect must consider the situation as it actually existed at the time the delay occurred.[198]

150 Can the client legally prevent the architect from giving an extension of time?

Under standard forms of building contract, it is the architect's duty to give an extension of time to the contractor if the relevant criteria have been met. The building contractor and the employer have entered into a contract whereby they both have agreed that when the contract criteria are satisfied, the architect will give the contractor an extension of time. Design and build contracts such as the JCT DB contract are excluded from this generalisation of course, because no architect is involved.

It is common for an architect to notify the employer before an extension is given, but it is not strictly necessary and it could be said to be misleading. The employer may think that the architect is seeking consent to the extension. If the architect does refer the extension of time to the employer, it is probably wise to make clear that the notification is a courtesy and for information only. Having said that, it is difficult to criticise an architect who takes the widest possible soundings before deciding on an extension of time. Not only the employer, but also the clerk of works and other consultants could be canvassed. Usually the architect will be seeking factual testimony so that the length of particular delays can be established with accuracy. The essential thing is that the architect must not only decide the extension of time, but also make that quite clear even though others may be consulted.

In *Argyropoulos & Pappa v Chain Compania Naviera SA*,[199] the (Claimant) architect, under the JCT Contract for Minor Building Works 1980, gave extensions of time to the contractor and the employer objected – even going to the extent of considering the extension of time clause 'no longer valid'. The employer refused to accept the extension and a later extension given by the architect. The architect was informed that employer's approval was required for any extension of time. At one stage the employer visited site and told

198 *Walter Lawrence v Commercial Union Properties* (1984) 4 Con LR 37.
199 (1990) 7-CLD-05-01.

the contractor that the architect had no power to give extensions of time. Eventually, the architect, on the advice of solicitors, withdrew its services. The extension of time point was just one part of the case, but in relation to that the judge said:

> . . . the Defendants sought to interfere with the [architects'] performance of their duties under [the extension of time clause] which they very properly resisted. Some of [the Defendants'] letters were also very offensive and indicated a total lack of confidence in the [architects]. [The Defendants and their] Solicitors also undermined the Plaintiffs' position in relation to the contractors. In my judgment the Defendants' letters, the Solicitors' letters and the Defendants' conduct were in breach of contract and the [architects] were amply justified in treating their engagement as at an end.

Not only does that show that interference with the architect's duty to give extensions of time is unlawful, in some circumstances, it probably amounts to repudiation on the part of the employer.

151 Must the architect give the contractor detailed reasons to explain the extension of time?

Most standard forms of contract require the contractor to give notice of delay to the Works as soon as it becomes reasonably apparent. For example, SBC clause 2.27 requires the contractor to notify the architect in writing of all such delays irrespective of whether or not they are relevant events. The clause continues to require the contractor to provide specify details of the delaying event and its likely effect and to provide the architect with such further information as reasonably required. Armed with this information, the architect must make an extension of time. SBC clause 2.28.3 sets out what the architect must include in the decision. It is very brief and amounts to notification of the extension of time attributed to each relevant event and the reduction in time (if any) attributed to each relevant omission. Therefore, the contractor will receive, in return for possibly several files of information, a single sheet of paper setting out the bare bones of the decision. Contractors often ask for more details. What they want are details of the way in which the architect has calculated the extension of time. They want to know precisely why some of their delay notices apparently have been ignored. At one time it was common for architects to send very detailed responses and to receive, in return, detailed objections from contractors. From the contractor's point of view, it is obvious why it wants to see the architect's reasoning. Only then can it start to build its case.

There are three very good reasons why an architect should not provide details of the reasoning behind extension of time.

- The contract does not require it.
- It will simply provoke a series of exchanges with the contractor regarding something about which the architect must make the final decision under the contract.
- The contractor will find it more difficult to challenge an extension of time if the underlying reasons are not known.

If this appears to be rather mean-spirited on the part of the architect, it must be remembered that the architect has to fix an extension of time which is fair and reasonable. The architect is called upon to exercise judgment and not to use a coldly logical approach.[200] If the contractor, or indeed the employer, wishes to challenge the extension of time, the adjudication procedure provides a quick mechanism. In that forum, the architect, if required by the adjudicator, must provide reasons for the extension of time and the adjudicator must decide whether the contractor has provided sufficient grounds to upset the architect's decision.

152 SBC: Is it permissible for the architect to give a further extension of time if documents from the contractor have not been received until after the end of the 12 weeks' review period?

Many commentators say that the architect is not bound by the 12 weeks' period because the period is not mandatory, but only directory on the authority of the Court of Appeal in *Temloc Ltd v Errill Properties Ltd*.[201] Commentators who take this view may perhaps not have read the judgment with the care it deserves. A careful reading shows that the court in *Temloc*, in saying that the 12 weeks was not a mandatory period, were actually interpreting the provisions against the employer that was seeking to rely upon them (the *contra proferentem* rule).

200 *City Inn Ltd v Shepherd Construction Ltd* [2008] BLR 269.
201 (1987) 39 BLR 30.

In that case, the employer had stated '£nil' as the figure for liquidated damages and the Court of Appeal held that, if the contractor did not complete the work until after the date for completion, liquidated damages would be chargeable only at the stipulated rate, which would amount to no liquidated damages at all. The court said that the employer could not decide to claim unliquidated damages instead. In the usual way, the contract provided that after practical completion the architect must, within 12 weeks, confirm the existing date for completion or fix a new date. The architect did not act within the 12 weeks and the employer's position was that the liquidated damages clause could be triggered only if the architect carried out the duty within the 12 weeks. Therefore, the employer asserted the right to claim unliquidated damages for breach of an implied term. It was in this context that the court said that the time period was not mandatory.

The court apparently accepted the architect as the employer's agent. The matter could have been cleared up very simply on the basis that if the employer's argument succeeded, it would have been contrary to the established principle that a party to a contract cannot take advantage of its own breach.[202]

In a more recent case, the 12-week review period was confirmed in the following paragraph: 'The process of considering and granting extensions of time is to be completed not later than 12 weeks after the date of practical completion . . .'.[203] Therefore, it follows that if the contractor submits information after the 12 weeks has expired, the architect cannot consider that information. That is the case even if it is clear that if the information had been provided on time an extension of time would have resulted. To avoid this kind of difficulty, or at any rate to avoid any uncertainty, it is good practice for the architect to write to the contractor shortly after practical completion with a reminder about the deadline and requesting any further information that the contractor wishes to be considered no later than, say, week 7 of the 12. That puts the contractor on notice and, if the information is not provided on time, the contractor has no one else to blame.[204]

Obviously, sometimes there are pressing reasons why it is desirable that the architect considers late submissions, for example when

202 *Alghussein Establishment v Eton College* [1988] 1 WLR 587 (HL).
203 *Cantrell v Wright & Fuller Ltd* (2003) 91 Con LR 97.
204 *London Borough of Merton v Stanley Hugh Leach Ltd* (1985) 32 BLR 51.

to do otherwise would be to risk time becoming at large. In these hopefully isolated cases, the parties to the contract can agree to give the architect power to consider the matter after the 12 weeks has expired. That is because the parties to a contract can always agree to vary the terms of their contract if they wish.

Chapter 19

Liquidated damages

153 Is there a time limit for the issue of the certificate of non-completion under SBC and IC?

The certificate of non-completion is governed by clauses 2.31 and 2.22 in SBC and IC respectively. These clauses provide that the certificate must be issued by the architect if the contractor fails to complete the Works by the contract date for completion or any extension of that date. There is no express stipulation that the certificate must be issued by any particular date, although it is surprising how many people believe that it must be issued within 7 days of the contractor's failure to complete. This is incorrect. The only time limit is that imposed by the issue of the final certificate. The final certificate is the architect's final action under the contract. After issuing it the architect is *functus officio* – that is to say the architect has no further powers or duties and, therefore, cannot issue the non-completion certificate.

It may be argued, with some merit, that the contract clearly envisages that the certificate will be issued promptly because, on a practical level, the later the certificate is issued – the less money will be available from which the employer can deduct liquidated damages.

154 The employer terminated in the 9th month of a 10-month contract. Can the employer deduct liquidated damages from the original contractor until practical completion is achieved by others?

In general terms it appears that, if the employment of the contractor is terminated, the obligations of both parties are at an end in so far as future performance is concerned.[205] This seems to be perfectly in accordance with good sense, because if the Works are completed by another contractor the original contractor can have no control over the completion. That is not to say that a party will avoid the payment of damages accrued up to the time of termination.[206]

The decision in *Re Yeadon Waterworks Co & Wright*[207] suggests that the courts will support a specific term in the contract that provides that in the event of termination of the employment of a contractor and the completion by another, damages could be deducted until the Works are completed. In that case, however, the Works were completed by the guarantor of the contractor, which was probably the deciding factor.

The JCT series of contracts provide for termination of the contractor's employment, following which the employer may engage another contractor to enter site and complete the Works. Such a clause was held to be incompatible with the right to liquidated damages in *British Glanzstoff Manufacturing Co Ltd v General Accident Fire & Life Assurance Corporation Ltd.*[208] If a contractor has left the site, wrongly thinking that the Works are complete, it seems that contractor will be liable for liquidated damages until the Works have in fact been completed by a replacement contractor.[209] The precise wording of the clause in the contract will be the deciding factor. In the New Zealand case of *Baylis v Mayor of the City of Wellington,*[210] liquidated damages were held to be deductible after termination, because the clause specifically excluded entitlement during the time taken by the employer to secure a replacement contractor.

In *Re White,*[211] the electric lighting contract contained what was held to be a liquidated damages clause. The court remarked that there was a clause in the contract which gave the engineer power, if necessary, to employ other contractors to complete the Works, and provided that the defaulting contractor should be liable for the loss so incurred without prejudice to his obligation to pay the liquidated damages under the contract. It is not clear from the report whether

205 *Suisse Atlantique etc v NV Rotterdamsche Kolen Centrale* [1966] 2 All ER 61.
206 *Re Morrish, ex paste Sir W Hart Dyke* (1882) 22 Ch D 410.
207 (1895) 72 LT 832.
208 [1913] AC 143.
209 *Williamson v Murdoch* [1912] WAR 54.
210 (1886) 4 NZLR 84.
211 (1901) 17 TLR 461.

the employer was seeking liquidated damages beyond the date of termination. The employer does not, however, appear to have claimed anything other than liquidated damages, despite the words of the contract, which appear to give the employer the right to claim liquidated damages for breach of obligation to complete on time until the date of actual completion, together with all the additional costs associated with completion by another contractor.

The effect of termination on the right to recover damages was considered in *Photo Production Ltd v Securicor Transport Ltd.*[212] Speaking of another case,[213] it was said:

> . . . that when in the context of a breach of contract one speaks of 'termination' what is meant is no more than that the innocent party or, in some cases, both parties are excused from further performance. Damages, in such cases, are then claimed under the contract, so that what reason in principle can there be for disregarding what the contract itself says about damages, whether it 'liquidates' them or limits them, or excludes them?

This seems to be a clear reinforcement of the view that there can be no continuing liability to pay liquidated damages, but damages already accrued are recoverable. Standard forms of building contract state the grounds on which either party may terminate the contractor's employment under the contract. Many of the grounds for termination under the provisions of the contract are not breaches that would entitle the employer to terminate save for the express provision. It is thought that an employer who terminated using the contract provisions is restricted to recovering the amounts stipulated in the contract.[214]

Current building contracts do not appear to allow the continued deduction of liquidated damages after termination. In any event, the circumstances set out in the question suggest that, even if the contractor's employment was not terminated, liquidated damages would not be due until a further month had passed, because at the date of termination the date for completion had not been reached.

212 [1980] 1 All ER 556.

213 *Harbutt's Plasticine Ltd v Wayne Tank and Pump Co Ltd* [1970] 1 All ER 225.

214 *Thomas Feather & Co (Bradford) Ltd v Keighley Corporation* (1953) 52 LGR 30.

155 Can an employer suffering no actual loss still deduct liquidated damages?

The whole idea of liquidated damages is that it is an agreed amount which the parties have agreed shall be paid on the occurrence of some event. In relation to construction contracts, the event is usually failure by the contractor to complete by the completion date specified in the contract. If there was no such agreement in construction contracts, the employer would be obliged to take legal action through the courts to recover any losses suffered as a result of the late completion. That would involve proving that the contractor had a contractual duty to complete by a certain date, that the contractor failed to complete and the amount of loss that the employer suffered as a direct result. To achieve that through the courts or even arbitration would be time-consuming and expensive. In order to avoid that situation the parties agree, and standard form construction contracts have special clauses, that an agreed sum will be payable in the event of late completion. The sum is usually expressed as per week or per day.

It is established that the sum must be a genuine pre-estimate of loss as viewed at the time the parties entered into the contract. In other words, it must be the employer's best estimate of the loss which would be suffered if the contractor delays completion. It does not matter if the likely loss is difficult to estimate so that the employer can only make an informed guess. Once the sum is in the contract and the contract is agreed, the employer may recover the sum if the contractor defaults. The employer is free to recover less than the amount stated, but not more. The employer does not have to prove the loss: that is the whole purpose of liquidated damages. Therefore, the employer may recover the whole sum for the whole period in which the contractor is in default of completion even if there is no loss or even if the employer makes a profit as a result of the late completion.[215]

156 Why do contractors sometimes say that the employer cannot deduct penalties?

The terms 'liquidated damages' and 'penalty' are often used as though they meant the same thing. Contractors often refer to the 'penalty clause'. They are, of course, entirely different things. Liquidated damages are intended to compensate the employer and

215 *Clydebank Engineering Co v Don Jose Yzquierdo y Castenada* [1905] AC 6.

they should be a genuine attempt to predict the damages likely to be suffered as a result of a particular breach (usually the contractor's failure to complete on time). On the other hand, a penalty is a sum which is not at all related to probable damages, but rather inserted as a threat, a deterrent or a punishment. The courts will enforce liquidated damages, but not a penalty.[216] It is, therefore, of great importance to establish whether a sum in the contract is liquidated damages or a penalty. The courts pay no regard to the terminology, nor indeed to whether the parties have agreed the sum, but only to its true nature.

There is a mass of old cases about this topic, but more recently the Court of Appeal set out four principles to differentiate liquidated damages from a penalty:[217]

1. The parties' intentions must be identified by examining the substance rather than the form of words used.
2. A sum would not be a penalty where a genuine pre-estimate of loss had been carried out.
3. The contract should be construed at the time the contract was made, not at the time of the breach.
4. It would be a penalty if the amount was extravagant or unconscionable compared to the greatest foreseeable loss.

The greatest danger is probably where the building is to completed in sections and a separate liquidated damages sum is to be inserted for each section. Care must be taken that each sum bears a suitable relationship to its section. The insertion of the same sum for each section may suggest that some or all such sums are penalties. It should be noted that it is comparatively unusual for a sum expressed in a contract as liquidated damages to be held by the courts to be penalties.

157 Is it true that, where there is a liquidated damages clause, by implication there must be a bonus clause in the same amount for early completion?

This is what may be termed a 'construction contract myth'. A contractor may sometimes try to argue this in order to avoid having to

216 *Watts, Watts & Co Ltd v Mitsui & Co Ltd* [1917] All ER Rep 501.
217 *Jeancharm Ltd v Barnet Football Club Ltd* (2003) 92 Con LR 26.

pay liquidated damages on the basis that, without a corresponding bonus clause the liquidated damages clause is invalid. There is no truth and no legal authority whatever in that view. The employer may decide to have a bonus clause, but that is entirely unconnected to a provision for liquidated damages and the amounts need not be related.

If a bonus clause is included, its structure will depend on the requirements of the employer. Often, such a clause will provide for a relatively modest payment if the contract completion date is beaten by a few days increasing to significantly larger sums if the contractor succeeds in achieving earlier completion dates. However, it should be remembered that if a bonus clause is inserted, the contractor will probably request information much earlier than usual on the basis that it needs it earlier if it is to earn the bonus. It will be difficult to resist this argument. It is difficult to see how the information release schedule in SBC can sit happily beside a bonus clause and some other clauses would require amendment also. There are some contracts, of course, which make special provision for bonus clauses.

158 In SBC, if an employer wants to be able to recover actual damages for late completion, is it sufficient that the relevant entry in the contract particulars has been filled in as NA (not applicable)?

Liquidated damages are inserted in standard building contracts to avoid the necessity for the employer to prove the damage suffered as a result of late completion. The parties agree that, if the contractor is late for reasons which do not entitle it to an extension of the contract period, the employer may recover or deduct a fixed sum for every day or week, as the case may be, by which the contractor exceeds the contractual completion date. If the employer wants to recover actual damages, that involves a reversion to the position which would exist if there was no liquidated damages provision.

Merely stating in the contract particulars that the rate of liquidated damages is NA would not remove the liquidated damages provision from the contract. In SBC the provision is contained in clause 2.32. Clause 2.32.2 refers to liquidated damages at the rate stated in the contract particulars. Therefore, if the rate of liquidated damages is stated as NA or even if all reference is removed from the contract particulars, this can only mean that the rate is not applicable. In other words, there is no amount of liquidated damages. Therefore, although the employer may be entitled to liquidated damages in principle, there is no amount to be charged. The contract is to be filled in by the contract administrator on behalf of the employer and all such entries will be interpreted *contra proferentem* (i.e. against the employer). There is still provision for liquidated damages, albeit amounting to £nil, and that is exhaustive of the employer's remedies for late completion.[218]

159 Can the employer still claim liquidated damages if possession of the Works has been taken?

Architects and contractors alike often labour under the mistaken impression that once the employer occupies the Works no further liquidated damages can be levied. Indeed, possession is often – wrongly – equated with practical completion. An examination of the definitions of practical completion given by the courts shows that the criteria are whether the Works, except for minor items, are complete and whether there are any visible defects. Whether the employer is or is not in occupation is not a factor.

The case of *Impresa Castelli SpA v Cola Holdings Ltd*[219] gives some useful guidance. The contract was the JCT Standard Form of Contract With Contractor's Design 1998. Disputes arose and the parties made no fewer than three separate variations to the contract terms. In particular, the liquidated damages amount was doubled and the date for completion was amended three times. Significantly, it was agreed that the employer (Cola) could have access to the hotel in order for it to be fully operational.

The contract finished late and Cola claimed liquidated damages of £1.2 million. Impresa challenged that amount, arguing that Cola had taken partial possession and, therefore, the liquidated damages should be considerably reduced. Significantly, the court decided that there was nothing to suggest that partial possession had occurred. It would have been quite simple to have referred to 'partial possession' in any of the three agreements if that is what had been intended and the situation would have been different (see question 179), but there

218 *Temloc Ltd v Errill Properties Ltd* (1987) 39 BLR 30.
219 (2002) 87 Con LR 123.

was no such reference. Instead, the court concluded that clause 23.3.2 (which was virtually the same as clause 2.5 of the current DB and clause 2.6 of SBC) had been operated, which allowed the employer to use and occupy the Works with the contractor's consent. The court decided that is a lesser form of physical presence on the site than possession. Therefore, the full amount of liquidated damages was recoverable by Cola.

Sometimes there is no agreement at all, but the employer, perhaps frustrated at continuing delays, simply decides to move in. The courts have held that such occupation did not preclude the deduction of liquidated damages.[220]

160 SBC: If practical completion is certified with a list of defects attached, can the employer deduct liquidated damages until termination (which occurred later due to the contractor's insolvency)?

Clause 2.32.2 of the contract specifies that the employer may give notice that payment of liquidated damages is required or that liquidated damages will be deducted from the date on which the contractor should have completed the Works (or any section) until the date of practical completion. Practical completion marks the date on which the contractor has completed its obligation to construct the Works and, therefore, the employer is no longer suffering any damage due to non-completion after that date.

Although, in *Tozer Kemsley & Milburn (Holdings) Ltd v J Jarvis & Sons Ltd & Others*,[221] the judge, without criticism, refers to a schedule of defects added to the certificate of practical completion, it is established that a certificate of practical completion cannot be issued if there are known defects in the Works.[222]

Therefore, it is clear that the certificate of practical completion should not have been issued while there were known defects in the Works. In this instance, it would probably be open to the employer to seek adjudication or arbitration on the basis that the practical completion certificate had been wrongly issued. If the adjudicator or arbitrator agreed, the employer would be entitled to recover liquidated damages until the date of practical completion as properly

220 *BFI Group of Companies Ltd v DCB Integration Systems Ltd* (1987) CILL 348.
221 (1983) 4 Con LR 24.
222 *Westminster Corporation v J Jarvis & Sons* (1970) 7 BLR 64.

certified. If termination due to the contractor's subsequent insolvency occurred after practical completion, it would be irrelevant so far as the calculation of liquidated damages was concerned. If the employer does not take this step, the certificate is the cut-off point for liquidated damages and, so far as the list of defects is concerned, it would be subsumed into defects occurring during the rectification period.[223]

161 If the employer tells the contractor that liquidated damages will not be deducted, can that decision be reversed?

More often than one might expect, an employer will tell a contractor that liquidated damages will not be deducted either totally or for part of the contractor's delay. This is usually done because a contractor is complaining bitterly about life in general and the contract in particular and the employer is worried that, if faced with the prospect of huge liquidated damages, the contractor may simply walk away from the project. The contractor would be wrong to do that and liable for damages, but completing the project with a fresh contractor and trying to recover damages from the old contractor would be time-consuming and traumatic.

Sometimes an employer has been known to state that liquidated damages will not be deducted instead of the architect giving an extension of time that is clearly due. That kind of conduct is asking for trouble. It derives from the false notion that, if the contractor is not given an extension of time, it cannot subsequently make a claim for loss and/or expense. The actual result is that time probably becomes at large and the contractor can still make an application for loss and/or expense if it is so minded. One thing is certain: the employer should not say that liquidated damages will not be deducted and then undergo a change of mind.

That is what happened in *London Borough of Lewisham v Shephard Hill Civil Engineering*.[224] In that case, the contract was the ICE 6th edition, but the principle holds good for JCT contracts also. The case was actually an appeal to the court from the decision of an arbitrator. It was alleged, before the arbitrator, that Lewisham or its

223 *William Tomkinson and Sons Ltd v The Parochial Church Council of St Michael* (1990) 6 Const LJ 319.

224 30 July 2001 unreported.

engineer had assured Shephard on several occasions that liquidated damages would not be recovered. Apparently these assurances were given orally and not confirmed in writing. At the time of the arbitration the liquidated damages stood at a considerable sum – about £550,000. Shephard contended that, as a result of these assurances, they had paid their sub-contractors without deducting damages which otherwise they would have deducted. The arbitrator noted that the engineer's final extension of time was not given until about 2 years after substantial completion of the Works and Lewisham did not claim the liquidated damages until after the arbitration had begun and was by way of a counterclaim.

In all the circumstances, the arbitrator concluded that Lewisham did make the representations about liquidated damages. The arbitrator, therefore, held that Lewisham was estopped (prevented) from claiming liquidated damages, the basis of that decision being the well-known principle that if one party to a contract makes a representation to the other that it will not enforce particular terms of the contract and if the other party relies on that representation to its detriment, the first party will be estopped from later trying to enforce that particular term. Shephard had relied on Lewisham's representation to its detriment when it had paid its sub-contractors without deduction. The court held that the arbitration had jurisdiction to decide the issue.

On the assumption that the representations were given as alleged, the decision was clearly correct. An important point was that Shephard had relied on the representation when deciding to pay its sub-contractors without deduction. That was a very clear-cut case of reliance. Even if payment of sub-contractors was not an issue, most contractors would be able to show that they had acted differently as a result of a representation not to deduct liquidated damages.

Chapter 20

Loss and/or expense

162 It is 3 months after practical completion and the contractor has just produced a claim in four lever arch files. What should be done about it?

Under clause 4.23 of SBC and clause 4.17 of IC, the contractor's entitlement to loss and/or expense is made subject to a written application having been made to the architect within a reasonable time of it becoming apparent that the contractor is incurring or is likely to incur direct loss and/or expense for the grounds laid down. SBC actually goes somewhat further and makes the requirement a proviso or condition.

What is a reasonable time? What is reasonable is always a difficult question to answer, because the standard (and correct) answer is that it depends on all the circumstances. So far as a claim submitted 3 months after practical completion is concerned, it appears at first sight as though that is not within a reasonable time.

The first thing to be done is to examine the claim to identify the grounds and to relate the grounds to the times when the contractor knew that direct loss and/or expense was being suffered. That should be relatively easy to do. For each item, the architect must work out the date when the contractor must have known that it was either incurring or likely to incur the loss and/or expense. The next point to establish for each item is whether the contractor did make an application to the architect. In this context, it is suggested that the making of an application can be given quite a broad interpretation. Therefore, it might be enough if the contractor merely refers to a future claim for loss and/or expense for the item in question. The contractor is possibly giving notice of delay for the purpose of extension of time and may say something like 'and we shall require associated loss and expense'. The fact that loss and/or expense is not associated with extension of time should be ignored for this purpose. Or the contractor might say that it has received the architect's instruction and that it is likely to cause loss and/or expense. The point is whether the architect was given warning or, to use the legal phraseology, 'put on notice' that a claim for loss and/or expense was likely.

The whole point of the requirement for the contractor to make application within a reasonable time is to give the architect the opportunity to take steps at a time when there was still the opportunity to do so. For example, the architect may well have decided to cancel an architect's instruction (if that was the cause of the problem). It may be that other measures could have been taken by the architect or the employer to reduce the amount of likely loss and/or expense. Perhaps most importantly, the architect could have taken steps to ensure that adequate records were kept of the contractor's work and times. It is important to understand this in order to decide whether the contractor acted promptly.

If a contractor knew immediately it received an architect's instruction that loss and/or expense would be incurred, the application should have been made immediately. If the contractor realised the problem only when it was in the process of carrying out the relevant work, that is the time when it would have been reasonable to make the application. The application must identify the reason for the application, including the circumstances and the contract clause on which the contractor relies. Further information can be sent later. The important thing is that sufficient information is given to the architect to enable the basic claim to be identified. There is no obligation for the contractor to calculate the amount of the claim at this or any other time.

The phrase 'within a reasonable time' does not give the contractor freedom to wait for several months until notifying the architect. A reasonable time is clearly as quickly as practicable having regard to the contractor's own knowledge. If the contractor is late in making application and appears to be outside the time frame envisaged by 'a reasonable time', it is suggested that the final test to be applied is whether the employer will be prejudiced by the late application. For example, an employer could be prejudiced if the lateness of the application made it difficult for the architect to ascertain the facts or in some other way was unable to operate the machinery of the claims clause correctly.

In the case of the claim submitted 3 months after practical completion, any items that have already been the subject of a timely application, even if brief, must be identified and be considered. So far as the others are concerned, the application was made at least 3 months after the disruptive or delaying events were known to the contractor. It is difficult to escape the conclusion that the architect has no power under the contract to consider them. That does not mean that the contractor has lost any remedy for these items. If they are also breaches of contract or other defaults, taking action at common law for damages is still available.[225] Of course, it is always open to the employer to agree with the contractor that, even if late, the architect will consider the application. In this instance, the architect must make sure to notify the employer in writing of the ways in which the employer may be prejudiced by the late application and obtain the employer's express instruction to proceed.

Once this initial point has been established, the architect can proceed to consider the claim in the usual way: by checking that the contractor has identified the contract clauses and relevant matters on which it relies; by establishing that the facts support the contention that there has been an occurrence that falls under the relevant matter; that the occurrence has resulted in direct loss and/or expense for the contractor; and that the contractor would not be reimbursed by any other payment under the contract. It should be noted that the contract does not refer to the contractor having been reimbursed, but only that it would have been reimbursed. Therefore, the point is not whether the contractor has been paid, but whether the contractor was entitled to be paid under another clause.

163 What exactly is a global claim?

A typical global claim is where a contractor puts in a claim for loss and/or expense along the following lines: 'The architect has issued 125 architect's instructions requiring variations and all those instructions resulted in a delay to the completion date of 35 days. The cost of the overrun is £XXX.' The difficulty facing the contractor is that if it can be shown that some of the architect's instructions had no effect on the completion date, the claim falls to the ground, because there is no mechanism for working out how much of

225 *London Borough of Merton v Stanley Hugh Leach Ltd* (1985) 32 BLR 51.

the total amount the contractor allowed for those architect's instructions.

When making a claim either at common law for damages or under a standard form contract for loss and/or expense, it is usually necessary for the contractor to prove each part of the claim, by relevant evidence. Therefore, in the example noted above, the contractor should separate each architect's instruction and provide evidence of the effect of each separate instruction on the completion date. Sometimes, a contractor will lump all elements of the claim together including both prolongation and disruption and that compounds the problem.

However, there are some circumstances in which a 'global' claim may be admissible. This principle also applies to extensions of time. The case which is usually cited in support of the global approach is *J Crosby & Sons Ltd v Portland UDC.*[226] It is a case decided under the ICE Conditions of Contract (4th Edition) which established the criteria for a global claim.

The court held that a global approach can be justified only in those circumstances where a claim depends 'on an extremely complex interaction in the consequences of various denials, suspensions and variations' and where 'it may well be difficult or even impossible to make an accurate apportionment of the total extra cost between the several causative events'. In those limited circumstances, the court said, there is no reason why an architect, engineer or arbitrator 'should not recognise the realities of the situation and make individual awards in respect of those parts of individual items of the claim which can be dealt with in isolation and a supplementary award in respect of the remainder of those claims as a composite whole'.

This does not, of course, relieve the contractor of producing substantiating evidence and proving each head of claim. What it does is to enable the architect or quantity surveyor to adopt a commonsense method of ascertaining certain complex claims where it is impossible or totally impracticable to prove the cost resulting from each individual item. The court went on to say:

> The events which are the subject of the claim must be complex and interact so that it is difficult if not impossible to make an accurate apportionment. It is very tempting to take the easy

226 (1967) 5 BLR 121.

course and to lump all the delaying events together in order to justify the total overrun or total financial shortfall. That argument is justifiable only if the alternative course is shown to be impracticable.

It is doubtful that the *Crosby* case can be relied upon in most cases. Contractors are faced with real difficulties if the facts are actually interconnected in such a complex way that it is not possible to separate them into different heads of claim. It is inequitable if an employer responsible for just one occurrence resulting in a claim, and for which a clear cause and effect can be demonstrated, is more likely to be made to suffer the consequences than an employer who has been the cause of many events, but all of which are inextricably bound together. Perhaps because of this, the courts have continued to allow claims to be made on a global basis.[227]

It may be sufficient if the contractor sets out the claim in enough detail so that the employer knows what is being claimed and, in many cases, it may be that the employer can quite readily calculate the amount the contractor should be paid without the necessity of the contractor having to jump through hoops in order to separate the claim into its various parts for the purpose of allocating a value to each part.

The basic position is that the contractor is entitled to put its claim in any rational way but, if the claim is presented on a global basis, there may be difficulties in providing the necessary evidence. The need to do that has been re-stated in an Australian case.[228] It held that where the connection between cause and loss is not otherwise apparent, each aspect of the connection must be set out unless the probable existence of such a connection can be shown by evidence or by argument, or unless it can be shown that it is impossible or impracticable for it to be itemised further.

The subject of global claims was examined again from basic principles by the Scottish courts in *John Doyle Construction Ltd v Laing Management (Scotland) Ltd*.[229] This judgment re-emphasises the

227 See for example *Nauru Phosphate Royalties Trust v Matthew Hall Mechanical and Electrical Engineers Pty Ltd and Another* (1992) 10 BCL 178; *Bernhard's Rugby Landscapes Ltd v Stockley Park Consortium Ltd* (1997) 82 BLR 39.

228 *John Holland Construction & Engineering Pty Ltd v Kvaerner RJ Brown Pty Ltd* (1997) 82 BLR 81.

229 [2002] BLR 393.

point that, for a global claim to succeed, the employer must be responsible for all the major causative factors. That is a point often overlooked when such claims are made.

There is nothing in the SBC, IC or ICD forms of contract which requires the architect to give the contractor the benefit of any doubt where claims are concerned. It is rather the contractor's task to prove, on the balance of probabilities, that what it asserts is correct. It is tempting for a contractor who is faced with substantial delays to try and secure at least some additional payment by making a very broad-based global claim but, unless the architect and quantity surveyor are both half asleep, the chances of such a claim succeeding are very slim.

164 How can a contractor claim for disruption?

Disruption has always been very difficult to establish with any precision and even more difficult to ascertain in monetary terms, which why the great majority of contractors' claims are based on prolongation of the contract period which is easier to understand and an easier claim to make and to deal with, although not if all parties are acting strictly in accordance with the terms of the particular standard form. Commonly, a contractor's claim for disruption relies on comparing anticipated with actual labour costs. This approach is lacking in any kind of merit and it has been very seriously criticised in the courts.[230] The inescapable facts are that there may be many reasons for the costs of labour being greater than the contractor anticipated at tender stage, other than reasons for which the employer or the architect can be held accountable.

An acceptable method of evaluating disruption is to compare the value to the contractor of the work done per man during a period of no disruption with the value per man doing the disrupted period and then to apply the ratio obtained to the total cost of labour.[231] It is obvious that, if this method is to work, it must be possible to isolate a period free from disruption. The comparison must relate to similar work.

Disruption often affects non-critical parts of a project, but not to the extent that those parts become critical in programming terms.

230 *London Borough of Merton v Stanley Hugh Leach Ltd* (1985) 32 BLR 51.
231 *Whittal Builders Co Ltd v Chester-Le-Street District Council* [1996] 12 Const LJ 356 (the first case).

That is why it is difficult to formulate a convincing claim. It is usually necessary to deal with each instance separately. The laying of specialist flooring in certain parts of a building may not be critical and there may be so much float that architect-induced delays do not cause the activities to become critical. Nevertheless, there may be serious costs involved as a result of the delay. In such instances, the contractor should provide its claim for each instance as a separate item, comparing the time it should have taken to do the work with the time actually taken.

165 Why are overheads and profit difficult to claim?

If a contractor was kept on site for longer than the contract period, it used to be accepted that the contractor would be able to recover overhead costs and loss of profit for the whole of the period of delay, provided that the delay was not due to its own fault of course. That view is no longer correct and the recovery of head-office overheads is now quite difficult. The usual basis of claim for recovery of head-office overheads is not that the contractor has actually lost the overhead sum, but that it has lost the opportunity to recover its overheads in the price of another contract by being kept on the particular site after the overheads allocated to that particular contract have been exhausted. The basis of the claim is that the contractor allocates a percentage of head-office overheads to each project. For example: if head-office overheads are £18,000 per annum and the contractor has three projects of the same contract price each lasting 12 months, it will want to recover £6,000 overheads from each project and it will add that amount to each price. If one of the contracts lasts 14 months instead of 12 months, the contractor will be looking to recover overheads for the extra 2 months which it could have earned if it had another project of the same size immediately following the first, i.e. £1,000.

It follows logically that the contractor must be able to show that it had other work which it could have done during the delay period, otherwise, in the absence of any delay, there would have been no chance of contribution from another contract during the period and, therefore, no loss.[232] There are two kinds of overhead costs.

232 *JF Finnegan v Sheffield City Council* (1988) 43 BLR 124.

Head-office overheads include not only costs of staff engaged upon individual contracts but also such general items as rent, rates, light, heating, cleaning, etc. and also clerical staff, telephonists, etc. and general costs such as stationery, office equipment, etc. It is important to distinguish between these two elements of overhead costs however calculated. One set of overhead costs is costs which are expended in any event: rates, electricity and the like. The other is managerial time which is directly allocable to the project and to no other.

On every contract, delay or disruption may lead to some increase in direct head-office administrative costs, relating not only to any period of delay but also to the involvement of staff engaged in dealing with the problems caused by disruption, for example contract managers spending more time in organising additional labour, revising construction programmes, securing staff, ordering additional materials, arranging plant hire and so on. If, however, they would not have been fully employed, if it was not for the delay, the contractor may face difficulties in recovering such costs which, it will be argued, it would have incurred as part of its head-office expenditure in any event. The courts have held:

> . . . it is for [the contractor] to demonstrate that he has suffered the loss which he is seeking to recover . . . it is for [the contractor] to demonstrate, in respect of the individuals whose time is claimed, that they spent extra time allocated to a particular contract. This proof must include the keeping of some form of record that the time was excessive, and that their attention was diverted in such a way that loss was incurred. It is important, in my view, that [the contractor] places some evidence before the court that there was other work available which, but for the delay, he would have secured, but which, in fact, he did not secure because of the delay; thus he is able to demonstrate that he would have recouped his overheads from those other contracts and thus, is entitled to an extra payment in respect of any delay period awarded in the instant contract.[233]

The problem in that case was that the delay was not sufficient to stop a building contractor of the size and standing of the contractor from tendering for other work. The recovery of head-office overheads as

233 *AMEC Building Ltd v Cadmus Investments Co Ltd* (1997) 13 Const LJ 50.

part of prolongation costs is likely to be difficult in future where large contractors are concerned. Indeed, it is always difficult for any contractor to show that it has been prevented from using its workforce on another project, because the current project is delayed. In modern construction practice, much if not all of the workforce will be sub-contracted and the types of operatives engaged during a period of delay at the end of a contract are finishing trades and not the early trades needed for a new project. The supervisors will often be finishing foremen.

The use of formulae for calculating head-office overheads and profit was not approved by the High Court in *Tate & Lyle Food and Distribution Co Ltd v Greater London Council*,[234] especially if other more accurate systems are available, but the contractor fails to take advantage of them. This case throws doubt on the legitimacy of charging a percentage to represent head-office or any managerial time spent as a result of delay or disruption, at least in the absence of specific proof. It was held that expenditure of managerial time in remedying an actionable wrong done to a trading company was claimable at common law as a head of 'special damage'. The actual claim failed because the company had kept no record of the amount of managerial time actually spent on remedying the wrong and, accordingly, there was no proof of the claim. The High Court refused to speculate on quantum by awarding a percentage of the total of the other items of the claim. In order to make a claim involving either overhead levels or profit levels (or both), it appears that actual overheads and profits must be identified – not merely theoretical or assumed levels. If 'direct loss and/or expense' is the equivalent of what is claimable as damages for breach of contract at common law, the common law principles must apply. That is not to say that no contractor can ever claim loss of overheads or profit, but it is not by any means as straightforward as once assumed.

166 Why do contractors use formulae for calculating claims?

It is clear from the answer to the previous question that it is not easy for a contractor to recover lost overheads and profit. Even where a contractor can make out a claim in principle it is by no means easy to

234 [1982] 1 WLR 149.

calculate the amount of overheads and profit in a way which will convince the architect or the quantity surveyor carrying out the ascertainment. Formulae have been applied by the courts, but not recently. They appear to give a relatively easy way of calculating overheads and profit without the necessity of proving the sum claimed in the usual way by giving an itemised breakdown. Formulae assume a healthy construction industry and that the contractor has finite resources so that, if it is delayed on one project, it will be unable to take on other work. During a recession that may not be the case. If the contractor's workload is not heavy, or if the contractor is substantial, it will have difficulty in showing that a delay caused it to lose the opportunity to carry out other work. But when the construction industry is buoyant or booming at the material time, a formula approach may be acceptable.[235] Double-recovery is a danger when using formulae, particularly in respect of directly engaged administrative staff. If some or all of the prolongation period is caused by complying with instructions to carry out additional work, the contractor will have recovered an appropriate proportion of overheads.

A formula will not be acceptable if there is a danger that it will overstate the actual loss to the contractor, and the formula should be backed up by supporting evidence, for example, the tender make-up, head-office and project records and accounts, showing actual and anticipated overheads before, during and after the period of delay. Any formula should be used with caution by the contractor and it should be treated suspiciously by the architect and quantity surveyor. The best-known formula is probably the Hudson formula, although it is not ideal. The courts have sometimes referred to the use of this formula when actually using another, slightly less well known but better, formula: the Emden formula. The Emden formula is as follows:

$$\frac{h}{100} \times \frac{c}{cp} \times pd$$

where *h* equals the head-office percentage arrived at by dividing the total overhead cost and profit of the contractor's organisation as a whole by the total turnover; *c* equals the Contract Sum in question,

235 *St Modwen Developments Ltd v Bowmer and Kirkland Ltd* (1996) 14 CLD-02-04.

cp is the contract period and *pd* equals the period of delay, the last two being calculated in the same units, e.g. weeks.

This formula can be useful as an approach where actual costs of head-office staff directly engaged upon the individual contract are not obtainable. In that case, the proportion of the contractor's overall overhead costs that can be shown from its accounts to be spent upon staff directly engaged on contracts can be substituted for the element *h* to obtain a rough and ready approximation of the cost of staff engaged on the particular contract during the period of delay. However, this approach does not make an allowance for the cost of greater staff involvement during the original contract period due to disruption.

167 What are 'interest and finance charges' which the contractor is trying to claim?

Under the direct loss and/or expense provisions of the JCT Forms and probably under similarly worded provisions in other standard forms, interest and finance charges form part of a claim for loss and/or expense. What it amounts to is the loss of interest that might have been earned if the money claimed had been invested; in other words, compensation for the loss of use of money. The basis of the interest and finance claim is that the contractor should be compensated for the fact that the loss and/or expense would have been incurred some time before ascertainment and certification by the architect. It is important to understand that the interest is not to be considered as interest on a debt but as a constituent part of the loss and/or expense claimed.[236] The courts have decided that finance charges should be calculated on a compound interest basis.[237]

The date at which such interest and finance charges start to run is a vexed question. The contractor may contend that it is the date on which it makes its application and the architect knows that there is a finance matter to consider. The architect may say that the operative date is the date on which sufficient information is available to enable the interest to be considered. It is sometimes argued that, whenever the architect may ascertain, the reality is that the contractor is bearing the financing charges of the whole amount which is ultimately

236 *FG Minter Ltd v Welsh Health Technical Services Organisation* (1980) 13 BLR 1 (CA).

237 *Rees & Kirby Ltd v Swansea City Council* (1985) 5 Con LR 34 (CA).

ascertained as due and, in the meantime, the employer has full use of the money. It is thought that the better view is that the period commences when the architect has received all the information he requires to ascertain. The matter is in the contractor's hands. Any delay between making application and providing full information is the responsibility of the contractor and that should be taken into account in deciding on the operative period for calculation of the interest.

168 Can a contractor recover the professional fees of a consultant engaged to prepare a claim?

Although contractors will often include the cost of preparing the claim within the claim itself, the contractor is not usually entitled to reimbursement for any costs it has incurred in preparing the claim. The reason for that is that under SBC, IC, ICD and DB the contractor is not required to prepare a claim in the sense of a fully reasoned document such as might be used as the case in arbitration proceedings. The contractor is merely required to make a written application to the architect, backed up by supporting information. That ought not to amount to more than an application, setting out the circumstances, citing the clause numbers, the relevant matters involved and linking the two together. That should be something that any contractor which has genuinely suffered loss and/or expense could do without too much difficulty although it will involve proper research into the occurrences giving rise to the losses. Therefore, fees paid to so-called claims specialists or to independent quantity surveyors or other professional advisers are not in principle allowable as a head of claim.

If a claim proceeds to arbitration or litigation, the contractor is entitled to claim its costs which will include the cost of getting the claim into the right form for arbitration. On a summons before a court to review taxation of costs of an arbitration which was settled during the hearing, the fees of a claims consultant for work carried out in preparing the contractor's case for arbitration (essentially, three schedules to the Points of Claim) were allowed as costs of a potential expert witness in the arbitration.[238]

238 *James Longley & Co Ltd v South West Regional Health Authority* (1984) 25 BLR 56.

The expenditure of managerial time in remedying an actionable wrong done to a trading company can properly form the subject matter of a claim for 'special damage' in an action at common law.[239] Therefore, it is possible that, in principle, there may be a claim for the cost of managerial time within the company spent on preparing a claim. Obviously, there could be no element of double recovery and such a claim should not be covered by any claim for head-office overheads. It is likely to be quite difficult for a contractor to successfully make such a claim and there would be substantial hurdles to cross. The case against recovery of the costs of employing a claims consultant is that such employment is unnecessary and, therefore, the costs does not amount to money necessarily expended and so recoverable. It is likely that a contractor trying to recover such costs would be put to proof that the particular circumstances were such as made the employment of a claims consultant essential.

169 Does the contractor have a duty to mitigate its loss?

If the contractor wishes to claim loss and/or expense or damages under any of the standard form contracts, it has a duty to mitigate its loss. By 'mitigate' is meant to 'lessen'. Therefore a contractor has a duty to do what it reasonably can to reduce the amount of loss. The principles of mitigation are straightforward:

- The contractor cannot recover damages as may result from the employer's actions if it would have been possible to avoid such damage by taking reasonable measures.
- The contractor cannot recover damages which have been avoided by taking measures greater than what might be considered reasonable.
- The contractor can recover the cost of taking reasonable measures to avoid or mitigate its potential damages.

The contractor is not obliged do everything possible. If that was the case, a successful claim for damages or loss and/or expense might be rare. The contractor need not do anything more than an ordinary prudent person in the course of his business would do.[240]

239 *Tate & Lyle Food and Distribution Co Ltd v Greater London Council* [1982] 1 WLR 149.

240 *London and South England Building Society v Stone* [1983] 1 WLR 1242.

Obviously, the contractor must have the right, in principle, to the damages, before mitigation is relevant. It is important to understand that a failure to mitigate will not give rise to a legal liability, but it will reduce the damages recoverable to what they would have been had mitigating measures been taken.

For example, if the architect gave an instruction which resulted in a number of operatives standing idle for a few days, the contractor would not be entitled simply to keep the operatives on site on full pay until the opportunity arose for full employment again. It would be obliged to make reasonable endeavours to use the operatives elsewhere either on that or another site. The costs incurred by the contractor until, acting reasonably, it was able to move the operatives and the costs of the move would be likely to be recoverable as a part of a claim. In practice, the courts would not examine the contractor's attempts to mitigate too critically. The contractor's costs would probably still be recoverable if he acted reasonably, even if the contractor's actions resulted in an increase in its loss.[241]

The extension of time clauses of many standard form contracts require the contractor to use its best endeavours to prevent delay occurring and to reduce the effects of a delay.

Where the contractor is claiming loss and/or expense under a standard form contract, it is for the employer to demonstrate that the claimant has failed to mitigate.[242] Under most standard forms that is something which the architect would handle. In doing so, the architect would be entitled to request reasonable substantiating information.

170 Is it permissible to claim increased costs by reference to national indices?

A contractor making application for loss and/or expense under one of the JCT standard form contracts will often submit what it believes to be a substantiating case. If the architect is not satisfied that the information provided is enough to reasonably enable an opinion to be formed about the validity of the claim, further information can be requested. Many architects get stuck when faced with the question of increased costs. There is nothing particularly mysterious about them and it is best to consider them from the basics. The contractor has

241 *Melachrino v Nicholl & Knight* [1920] 1 KB 693.
242 *Garnac Grain Co Inc v Faure & Fairclough* [1968] AC 1130.

calculated its tender on the basis that it will be able to get on with the Works in accordance with the programme. The price of all materials at the time they will be required and the cost of all labour resources and equipment is taken in account. If the contractor suffers additional expenditure on labour, materials or plant due to increases in cost during a period of delay, it is an allowable head of claim. Claims on this basis may also be sustainable where disruption has resulted in labour-intensive work being delayed and carried out during a period after an increased wage award. Where a claim of this kind is being made in respect of a delay in completion, it is not only the period of delay that should be considered. The correct calculation would be the difference between what the contractor would have spent on labour, materials and plant and what he has actually had to spend over the whole period of the work as a result of the delay and disruption concerned. However, when making this calculation, proper allowance must be made for the recovery of any increased costs under the relevant fluctuations clauses in the contract, if applicable.

A contractor will often try to simplify such calculations by the use of some kind of formula or notional percentage, such as a national index, to produce a result. Such an approach is not acceptable, because the contractor must show that the increases in costs have been the consequence of the cited occurrence and the precise amount of loss and/or expense sustained thereby must be calculated. This can become quite complicated in the case of materials, because the contractor must show that it could not reasonably have placed its order earlier in order to avoid the increases. The proper calculation of increased costs sufficient to support a claim is not easy. Indeed, it is rather a tedious exercise. Moreover, it must not be assumed that all work and all materials after the period of delay or during a prolongation period after the contract completion date, will automatically suffer a price increase. That is again a tempting shortcut which the architect has a duty to the employer to reject.

171 The contractor is demanding to be paid 'prelims' on the extension of time. How is that calculated?

None of the standard forms of contract entitle the contractor to any payment as a consequence of extension of time. The purpose of extension of time clauses is that the period of time available for carrying out the contract Works can be extended. The contractor will look in vain for any reference in the contract to money in connection with extension of time. In short, there is no connection between the contractor's entitlement to an extension of time and any entitlement to loss and/or expense.[243]

There is no provision for the payment of 'prelims' in any of the standard forms. 'Prelims' is short for 'preliminaries' i.e. the first part of the bills of quantities or specification. When a contractor talks about recovering its preliminaries it means the price for the preliminary items that has been inserted in the bills or specification as part of the tender price. This preliminaries price is often inserted as a lump sum or as a price per week. Obviously a lump sum can easily be converted into a weekly rate, by dividing it by the number of contract weeks.

It was once very common, but now fortunately less so, for the architect to give a contractor an extension of time and then for the quantity surveyor to ascertain the amount payable to the contractor by multiplying the weekly amount of preliminaries in the specification by the number of weeks' extension of time. As construction professionals better understand their responsibilities in dealing with loss and/or expense, this practice is less common. It borders on negligence unless sanctioned by the employer in the full knowledge of all its implications. The contractor can never be entitled to recover its preliminaries as loss and/or expense, much less as a rate per week of extension of time, because the contractor is entitled to recover only its actual losses or actual expenses – in other words, the amount it can prove it has actually lost or spent.[244] It is not entitled to recover some notional amount, nor the amount inserted as part of its tender price which may, but more likely may not, be the same as its actual costs.

Therefore, the answer to this question is:

1 The contractor is not entitled to any money as a consequence of being given an extension of time.
2 Therefore, there is no need to calculate it.

243 *H Fairweather & Co Ltd v London Borough of Wandsworth* (1987) 39 BLR 106; *Methodist Homes Association Ltd v Messrs Scott & McIntosh*, 2 May 1997 unreported.

244 *FG Minter Ltd v Welsh Health Technical Services Organisation* (1979) 11 BLR 1; (1980) 13 BLR 7 (CA).

3 If the contractor makes a valid application under the claims clause (e.g. clause 4.23 of SBC or clauses 4.17 and 4.18 of IC), it is entitled to the actual amount of loss and/or expense it has suffered.

172 Is it true that a contractor cannot make a loss and/or expense claim under MW?

Strictly speaking, that is correct. The only clause in MW that mentions loss and/or expense is clause 3.6.3. It provides that if the architect issues an instruction requiring an addition to or omission from or any other change in the Works or the order or the period in which they are to be carried out and there is a failure to agree a price before the contractor carries out the instruction, it must be valued by the architect.

The architect is required to value it on a fair and reasonable basis using any prices in the priced document. Significantly, the valuation must include any direct loss and/or expense that the contractor has incurred as a result of regular progress of the Works being affected in either of two ways. The first is compliance by the contractor with the instruction. The second is the employer complying or failing to comply with clause 3.9.

Clause 3.9 requires both parties to comply with the CDM Regulations. In particular, under clause 3.9.1, the employer must ensure that the duties of the CDM co-ordinator are properly carried out and, if the contractor is not acting as the principal contractor, that the duties of the principal contractor are also properly carried out. The grounds on which the architect can include loss and/or expense are obviously quite restricted.

It is clear from the wording of clause 3.6 that the architect's inclusion of loss and/or expense in the valuation does not depend on any application by the contractor. Indeed, the only time the contractor is expressly required to provide information to the architect is under clause 4.8.1, where the contractor must provide all documentation reasonably required for computation of the final certificate.

In calculating the valuation, the architect will no doubt ask the contractor for information. Indeed, in practice most contractors will provide information in the form of an application for payment on a monthly basis. Although the contract does not preclude such applications, it does not confer any status upon them. The architect may take notice of or ignore the information as the architect deems appropriate, because the only factor the architect needs to take into account is the priced document, whether that is a priced specification, work schedules or a schedule of rates. It is entirely a matter for the architect how the loss and/or expense is calculated. Many architects link the amount to the length of any extension of time that has been given on account of architect's instructions. Although one can see some logic in this approach, there is no justification for arriving at the loss and/or expense by multiplying the number of weeks by the amount the contractor has inserted in the priced document as its weekly preliminaries cost. Loss and/or expense is the equivalent of damages at common law. As such, the damages must be proved; the architect must secure the necessary evidence to show how much loss the contractor has actually incurred. At best, the preliminaries figure in the priced document is the contractor's best estimate at tender stage. It may be an under- or an overestimate. It certainly will not represent actual costs.

The contractor cannot make a claim for loss and/or expense under the terms of MW, because they contain no mechanism to enable it to do so. MW does not have the equivalent of clauses 4.23 and 4.17 of SBC or IC respectively. Therefore, the contractor cannot make a claim under the contract for loss and/or expense for information received late. All is not lost, however. There is absolutely nothing to stop the contractor making a claim at common law in such cases, basing the claim on a breach of contract by the employer and claiming damages. Such action has received judicial blessing.[245] The architect cannot deal with such claims, because they are outside the contract machinery. They must be handled by the employer. If the architect receives such a claim, it must be forwarded to the employer immediately. The architect should refrain from expressing any view about the claim unless consulted by the employer. Theoretically, the employer should deal with the matter by separate legal advice and, if appropriate, pay the money directly to the contractor without an architect's certificate.

245 *London Borough of Merton v Stanley Hugh Leach Ltd* (1985) 32 BLR 51.

Chapter 21

Sectional completion

173 The contract is SBC, which includes provision for sections. The employer wants to rearrange the sections. Can that be done with an architect's instruction?

The fact that the contract is to be carried out in sections has been agreed between the employer and the contractor when they executed the contract. In other words, the sections are a term of the contract. Therefore, in order to change the sections it is necessary to have a further agreement between employer and contractor. It is not something the employer can unilaterally decide, any more than the employer can unilaterally decide to reduce the contract sum by 20 per cent.

Still less can the sections be changed by the architect through the medium of an architect's instruction. Apart from any other consideration, the architect is not empowered by the contract to issue an instruction to that effect. Therefore, any such instruction would be void.

If the employer wishes to rearrange the sections, the contractor's consent must be sought. Even where both parties agree, the change cannot be achieved by an architect's instruction. If the contractor is willing, the employer must organise the drafting of a special addendum to the contract setting out the variation agreed between the parties and any other matters that arise from the change (for example, it will be necessary to amend the liquidated damages). Both parties must sign the addendum.

174 The contract is SBC in sections. The dates for possession and completion have been inserted for each section. Section 2 cannot start until section 1 is finished. Is it true that possession of section 2 must be given on the due date even if it is the contractor's own fault that section 1 is not finished?

This is a common problem when the contract is divided into sections, each with its own date for possession and completion, but two or more of the sections are interdependent. For example, a refurbishment project may be divided into three sections, but section 2 may be dependent on section 1 in a practical sense, because the contractor cannot physically be given possession of section 2 until section 1 has been completed. That is usually because the occupants of section 2 have to be moved to section 1 when it is finished. Usually, the dates for possession and completion of each section are inserted into the contract as a series of dates. The date for completion of section 1 and the date for possession of section 2 will probably be separated by a week or so to allow occupants and furniture to be moved from one section to another.

If section 1 is not finished by the completion date, even due to the contractor's fault, the contractor is still entitled to take possession of section 2 on the appointed date in the contract particulars (see clause 2.4 of SBC). If it is physically impossible for such possession to take place, the employer will be in breach of contract. The contractor is, therefore, correct.

Where a project is split into sections, any extensions of time must be given in respect of the particular section affected by the delaying event. There is no provision that the delay in one section will affect another. Therefore, even if the whole of the delay to section 1 entitles the contractor to an extension of time, it will be only section 1 that is extended and not section 2.

If the cause of the delay is entirely the fault of the contractor, the architect may say that the contractor, being responsible for the delay to section 1, is clearly responsible for the delay to section 2 also and it cannot expect to take possession of section 2 on the date set out in the contract particulars. This approach is very common, but wrong. The cause of the delay to possession of section 2 is not the contractor's delay to section 1, but the fact that the two sections are linked. If they were not linked, the contractor's delay to section 1 would not affect section 2. One of the difficulties is that, where the dates for possession and completion are simply set out as a series of dates, there is nothing to warn the contractor about the likely problem.

There are two probable supplementary issues: the first is whether anything can be done to avoid the problem in new contracts; the second is whether anything can be done where the situation outlined in the question is currently in place.

To avoid the problem is relatively straightforward. The employer must clearly show the links in the sections. The contract particulars should be amended to delete the current setting out against 'Sections: Dates of Possession of sections' and in its place or on a separate, but properly attached and signed, sheet, section 1 would have a date for possession and a date for completion, but section 2 would not have a date for possession. It would simply state: 'The date for possession is x days after the date of practical completion of section 1.' Therefore, a delay to completion in section 1 (from whatever cause) would be reflected in the date of possession of section 2 and there would be no breach of contract, because section 2 could be given to the contractor on the due date. The date for completion of section 2 would not be inserted, but rather: 'The date for completion is x weeks after the date that possession of this section was taken by the contractor'. It is difficult to see the grounds on which the contractor could make any financial claim on the employer for delays to section 1 that cause a delay to the possession of section 2 if this method of setting out the dates was implemented. It has been said that the dates for possession and completion cannot be entered in this way, because they are not actual dates. That would be to take the wording too literally. The important thing is that the wording enables the dates to be unerringly calculated, albeit not until practical completion of section 1 has taken place.

How to rectify the situation if proper provision has not been made is slightly more complex. If there is provision for the employer to defer possession of any of the sections by the appropriate amount, the employer must do so and the contractor will be entitled to an extension of time and probably whatever amount of loss it has suffered as a result of the deferment of possession. If there is no deferment provision or if the delay exceeds the period of deferment allowed under the contract, the situation appears to be that there is a breach of contract which, dependent upon circumstances, may be a repudiation. The contractor would be entitled to recover as damages the amount of loss he has suffered. If there was no provision for the delay situation in the contract, the architect would be unable to make any extension of time and the contractor's obligation with regard to section 2 would be to complete within a reasonable time. Therefore, liquidated damages for this section would not be recoverable.

SBC and IC now provide for such breaches to be dealt with by extension of time and loss and/or expense under SBC clauses 2.29.6 and 4.24.5 and IC clauses 2.20.6 and 4.18.5 respectively. The amount payable to the contractor, whether by virtue of a loss and/or expense clause in the contract or as damages for the breach, may not be substantial. The contractor would have to demonstrate a loss and the situation is simply that section 2 has been pushed back in time.

175 If the architect gives an extension of time for section 1 and if all the sections have dates for possession which depend upon practical completion of the earlier section, is the architect obliged to give a similar extension of time for each section?

If sections are linked so that the date for possession of each section depends on the date of practical completion of the earlier section, difficulties can arise. Some of the problems arising from linked sections have been explored in the previous question together with some ways of dealing with them.

The problems were placed before the court in a recent case.[246] In that case, the parties contracted under the JCT 98 form of contract. The date for possession of the first section was fixed but the date for possession of each following section depended on the date of practical completion of the previous section. However, and this is what gave the real potential for difficulty, the completion dates for each section were fixed. Consequently, when the contractor was in delay for 8 weeks in the first section, the architect gave 4 weeks, extension of time and 4 weeks, extension for each following section. Therefore, there was a period of culpable delay to the first section of 4 weeks for which the employer deducted liquidated damages amounting to £48,000. The problem for the contractor was that it then found itself in a period of culpable delay for each succeeding section, because the date for possession was delayed by up to 8 weeks each time, but only 4 weeks extension of time had been given in each case. The

246 *Liberty Mercian Ltd v Dean & Dyball Construction Ltd* [2008] EWHC 2617 (TCC).

contractor argued that for each section after the first section, the extension of time should reflect the initial 8 weeks, delay to the first section which irretrievably fixed the date of possession of the following sections. The court decided that liquidated damages were recoverable for all the weeks of culpable delay in each section. The court referred to what it called the 'critical feature of this contract':

> . . . namely that the building works were always going to be carried out sequentially, and that the work on one section could not start until the work on the previous section had reached practical completion or (in certain instances) the stage of completion identified in the sectional completion schedule. It is plain from that schedule, and from the sectional completion agreement as a whole, that both sides were aware that culpable delay of 4 weeks on section 1 would automatically mean that work on sections 2, 3, 4 and 5 would start 4 weeks late.

In an old case about a project to be completed in sections which was heard by the House of Lords, a hospital was to be built in three phases or sections.[247] There was a provision that the third section was to commence 6 months after the certificate of practical completion of section one was issued, but that it must be completed by a specified date. That was the root of the problem, because section one was in delay, which delayed the issue of the certificate of practical completion and thus delayed the commencement of section three. But as the completion date for section three was fixed, the period available for the contractor to carry out the section three work was effectively reduced from 30 months to 16 months. There was provision in the contract for nomination of certain sub-contractors and the contractor requested the employer to nominate sub-contractors who could carry out their work within the reduced period. The employer was unable to do so, but argued that there must be an implied term that the section three completion date must be extended by the amount of any extension of time given to the contractor in respect of the section one delay. The court refused to imply a term such as the one requested. That meant that the parties would have to negotiate an extension to the completion date of section three with a

247 *Trollope & Colls Ltd v North West Metropolitan Regional Hospital Board* (1973) 9 BLR 60.

resulting increase in the contract sum. Clearly, the unfortunate result was not foreseen by the parties and, in this instance, it resulted in addition cost to the employer. Had the delay not been as long, the result might have been a shortened work period for section three. The court in the *Liberty Mercian* case does not seem to have been referred to this earlier case.

Chapter 22

Practical completion and partial possession

176 If the architect has issued a certificate of practical completion with 150 defective items listed and the contractor is not remedying them within a reasonable time, what can be done about it?

The leading case on the requirements for practical completion[248] states that practical completion cannot be certified if there are known defects in the Works. Therefore, a certificate issued with 150 defective items listed is, on its face, void or, perhaps more accurately, voidable if either party applies to an adjudicator. An adjudicator ought to find that the certificate was not properly issued and the contractor would be obliged to rectify the defects before a certificate could be properly issued.

If neither party seeks adjudication on the matter, and in practice the contractor is unlikely to do so because the certificate carries various advantages, the defects will become part of the defects to be dealt with during the rectification period.

If the contractor does not rectify within a reasonable time, the architect may issue an instruction requiring the defects to be made good followed by a compliance notice under SBC clause 3.11 or IC clause 3.9. If the contractor does not comply within 7 days, the employer may engage others and all the additional costs incurred by the employer will be deducted from the contract sum. Alternatively, if the employer is content to wait until after the end of the rectification period, the defects can be added to the schedule of defects and dealt with in the usual way.

248 *Westminster Corporation v J Jarvis & Sons* (1970) 7 BLR 64.

177 Is the contractor entitled to a certificate of practical completion after termination?

The certificate of practical completion indicates that the contract Works are almost complete, that there are no known defects and there are only minor things to be done. When the employment of the contractor is terminated under the contract, whether by the employer or by the contractor, the Works will never reach practical completion under that particular contract. The only way to complete the Works will be for the employer to enter into a new contract with another contractor. The new contract will not be for the Works as included in the original contract, but only for the balance of the Works.

When the new contract is completed, the architect will issue a certificate of practical completion for the Works included in the new contract. The certificate will be issued to the employer with a copy to the new contractor and it will refer only to the balance of the original Works. The original contractor will not receive a copy of this certificate, because it was not concerned in the new contract. The original contractor will not receive a copy of a certificate of practical completion of the Works in the original contract, because they were never completed.

It is surprising how often architects and contractors get confused about this.

178 Is there such a thing as 'beneficial occupation' and is the architect obliged to certify practical completion if the employer takes possession of the Works?

A contractor may argue that, if the employer takes possession of a project before the architect certifies practical completion, the employer cannot recover liquidated damages for the subsequent period until practical completion is certified. It is said that the employer has the benefit of occupying the premises (hence 'beneficial occupation') and the whole purpose of liquidated damages is to compensate the employer if occupation cannot be achieved by the due date. There appears to be no basis in law for that view. There is case law to suggest that, where there is no express provision for partial possession, liquidated damages may be recovered even though the employer has occupied the Works provided that practical completion has not been certified.[249]

Some standard form contracts (for example SBC clause 2.6) permit an employer to use and/or occupy part or all of the Works if the contractor consents, but this occupation does not affect the contractor's exclusive possession of the Works, nor does it affect the contractor's obligations or entitlements with regard to liquidated damages.[250] The position is likely to be different if there is provision for partial possession in the contract and the employer, with the contractor's consent, has taken partial possession of the whole of the Works.

179 Can the employer take partial possession of the whole building so that the architect need not certify practical completion?

The answer to this question depends on the terms of the building contract. Under SBC, DB, IC and ICD the employer may take partial possession of any part of the Works if the contractor agrees, and once partial possession has been taken, practical completion is deemed to have occurred in respect of those parts. It is unusual, to say the least, for partial possession to be taken of the whole of the Works. It may sometimes happen when the employer is anxious to move in to the building, but when the criteria for practical completion have not been satisfied and, therefore, the architect cannot issue a certificate to that effect. It has been held that if partial possession is taken of the whole of the Works, deemed practical completion of the whole of the Works occurs.[251]

The effect of that is that regular interim payments end, liability for liquidated damages ends, the contractor's obligation to insure the Works ends, the rectification period commences and the first half of the retention must be released. All this follows just as surely as if the architect had been able to certify practical completion. Presumably, at some later date, the criteria for the issue of a certificate of practical completion will be satisfied. It may seem otiose for the architect

249 *BFI Group of Companies v DCB Integration Systems Ltd* (1987) CILL 348; *Herbert Construction UK Ltd v Atlantic Estates plc* (1993) 70 BLR 46.

250 *Impresa Castelli SpA v Cola Holdings* (2002) 87 Con LR 123.

251 *Skanska Construction (Regions) Ltd v Anglo-Amsterdam Corporation Ltd* (2002) 84 Con LR 100.

to certify at a later stage, because all the practical consequences of practical completion have already been set in motion. However, under the contract, that is what the architect is obliged to do and there is nothing in the contract which removes this duty once the prescribed criteria have been satisfied. The subsequent issue of the certificate by the architect does have one very useful function. It records the date on which, in the architect's opinion, practical completion occurred. Alternatively, if the architect decides not to issue such certificate, it is a breach of the architect's obligations under the contract, but it is difficult to see what if any consequences flow from it. The architect could easily record his or her view on when practical completion would have been certified by sending a letter to the employer.

Chapter 23

Termination

180 The contractor is running over time. The architect has over-certified. Are there any problems if the employer wishes to terminate?

An architect who certifies more than the amount properly due to the contractor is negligent and in breach of his or her contractual obligations to the employer by which the architect undertook to administer the contract. All the standard forms of contract permit the architect to certify only those amounts properly due. Because payment certificates are cumulative, the architect should be able to retrieve the situation in the next certificate. Obviously, that would not be the case if the employer terminated, the contractor went into liquidation, or the certificate in which the over-certification took place was the one issued after practical completion. Interim certificates are not intended to be precisely accurate. It has been well said that they are essentially a means by which the contractor is assured of some cash flow that roughly approximates to the work carried out.[252]

It is doubtful whether the employer would have grounds to terminate the contractor's employment simply because the contractor was running over time. Under JCT contracts, the architect would have to be satisfied that the contractor was failing to proceed regularly and diligently. However, for the sake of this question let us suppose that there are adequate grounds to terminate. The architect will be obliged to confirm that to the employer. Then the architect has no option but to inform the employer that there has been some

252 *Sutcliffe v Chippendale and Edmondson* (1971) 18 BLR 149; *Sutcliffe v Thackrah* [1974] 1 All ER 859.

over-certification and that it would be better to wait until the work on site has caught up with the amount certified. How long that will be will depend on the degree of over-certification and the rate at which the contractor is currently working.

Of course, the employer can always go through the termination procedure and then reclaim the over-certified amount, together with any other balance due, after the Works have been completed by others. In doing so, however, the employer would have to challenge the architect's certificate, probably in adjudication. This would be additional cost which, if the adjudicator found in favour of the employer, the employer would assuredly try to recover from the architect.

In practice, instances of over-certification tend to be unusual and concerning relatively small amounts, usually the result of the valuation of defective work. It is the architect's responsibility to notify the quantity surveyor of all defects so that they will not be valued. In any event, certification is entirely the province and responsibility of the architect.[253] That is why architects should not simply accept valuations from quantity surveyors and blindly transfer the valuation figures to the certificate; they should ask for a simple breakdown of the valuation. It is not the architect's job to do the valuation again, but the architect should have enough information to be satisfied that the valuation looks about right.

181 Termination took place under SBC due to the contractor's insolvency. Can the liquidator insist that full payment of any balance plus retention is immediately payable?

Termination due to the contractor's insolvency is covered by clause 8.5 of SBC. The employer may terminate at any time by written notice. The interesting point is that, under clause 8.5.3.1, as soon as the contractor becomes insolvent, even if no written notice has been given by the employer, the provisions of the contract that require any further payment or release of retention cease to apply.

That means that, although the employer may be slow in taking action to terminate the contractor's employment, the employer's duty to pay is at an end except as set out in clauses 8.7.4, 8.7.5 and

253 *R B Burden Ltd v Swansea Corporation* [1957] 3 All ER 243.

8.8. It should be noted, however, that there is no longer any provision for automatic termination.

These clauses stipulate that, within a reasonable time after the completion of the Works by another contractor and of the making good of defects, an account must be drawn up of whatever balance may be due from employer to contractor or vice versa as the case may be after taking into account all the costs of finishing off including any direct loss or damage caused to the employer by the termination. If the employer decides not to complete the Works using others, a statement of account must be drawn up and sent to the contractor after the expiry of 6 months from termination.

It is not unknown for liquidators, either directly or more usually through the services of specialist insolvency surveyors, to threaten employers with proceedings if payment of all money is not made immediately. This contract now makes clear that the normal payment provisions are at an end. Therefore, if a certificate has already been issued, the employer no longer has any obligation to pay the sum certified. If a certificate is due, the architect no longer has a duty to certify. The employer or the architect should respond to the liquidator or the surveyor, referring to the contract clause as the reason why no further payment will be made until the final statement of account is prepared.

In these circumstances, the employer will often have delayed payment of a certificate because it is suspected that the contractor is about to become insolvent. Indeed, the employer's failure to pay and the failure of others may be the very reason why the contractor eventually becomes insolvent. Therefore, at the date of insolvency, not only is a certified amount outstanding, but it may have been outstanding for a considerable time. Is the employer bound to pay such a sum on the basis that the employer cannot take advantage of its own breach of contract (failing to pay a certified amount on time)? The answer to the question seems to be that the employer is bound to pay monies that were outstanding at the date of the insolvency. The use of the words '*which require further payment*' appears to support that, because a payment that is overdue is not a '*further payment*', but rather a payment that ought to have been made already. The point is not beyond doubt and, in practice, the courts may be reluctant to order such payment when there is no realistic chance of it being recovered later if the cost of completion of the Works proves to be more than the cash still retained by the employer.

182 The contractor has gone into liquidation and another contractor is needed to finish the project. The MD of the original firm has now formed a new company and is asking to be considered for the completion work. Is that a problem?

This is not an unusual situation. The important thing to remember is that a limited company is a legal entity which is quite separate from the individuals who are the shareholders or directors of the company. Therefore, the original contractor in liquidation, which is obviously a limited company (individuals and partnerships do not liquidate, they become bankrupt), is not the same as the new company formed by the former MD.

When completing a project by employing another company, it is usually essential that the employer does so only after seeking competitive prices from at least three contractors. That is certainly the case if the employer is hoping to recover some of the additional costs from the original contractor. Although that may be problematical in the case of insolvency, the employer presumably still will need to prove entitlement to use money already withheld. Therefore, even if considered, the new company would be only one of three companies tendering. However, before including the new company on the tender list, the architect should bear the following in mind:

- All the contractors on the list should be firms about which the architect has made enquiries as to their stability, financial situation and workmanship on previous projects.
- The architect owes a duty to his or her client to advise them about obvious pitfalls, albeit the client is the person who has to make the final decision.
- The new company has no track record of any kind, because it has just been formed.
- The track record of the MD is not the same as the previous company, but as MD of a company which has just gone into liquidation, it does not suggest that the new company is in good hands.
- Usually, it is important to obtain the names of referees and to requests reports about projects the company has completed for them. That will not be possible in this instance.
- Are there any other circumstances which would reduce the risk for the employer?

In short, an architect who advises the client to include the new company in the tender list may be negligent.

183 If the contractor discovers that it has under-priced a project and cannot afford to carry on, must the architect try to negotiate an amicable termination and settlement?

A contractor may discover the under-pricing at various stages. If it is discovered immediately the tender has been submitted or in any event before the tender has been accepted, the contractor can simply withdraw the tender. This cannot be done if the employer has paid each tenderer a sum of money in return for keeping each tender open for a fixed period. Where an invitation to tender, or even the tender itself, merely states that the tender remains open for acceptance for a period, it will not preclude the contractor from withdrawing the tender, unless the employer has paid for the extended period. Even if the employer has paid for an extended period during which the tender remains open, the tenderer can still write to the employer setting out the position and pointing out that it cannot carry out the contract. It is difficult to imagine an employer proceeding to accept the tender in those circumstances, knowing that the contractor could not carry out its side of the bargain. There could be no valid contract, because there would be no true agreement, because the employer would have known the position before attempting to accept the contractor's tender.

Once the contract has started on site, the position becomes very much more awkward to disentangle. One might nurse some faint hope that, where a contractor suddenly discovered a massive under-pricing when the project was 25% advanced, the parties (albeit not legally obliged to do so) would sit down and either agree some further payment from the employer or agree that the contractor could simply leave the Works on payment of an agreed sum in nominal damages. Strictly, of course, once the tender has been validly accepted, there is a binding contract in existence between the parties and if the contractor is unable to proceed due to under-pricing, it is nothing more or less than repudiation of the contract for which the employer, who would have little choice but to accept, can claim the damages suffered. Such damage would presumably amount to the difference between the (albeit low) price accepted from the first contractor and the higher price for which a second contractor would agree to complete the Works.

Where does the architect fit into this scenario? Strictly, the architect has no role in sorting out the mess. That is assuming that the architect was not responsible for allowing the employer to accept a price which was obviously too low. If the architect did everything that could be expected so far as checking the tenders was concerned, the decision to accept the tender was the employer's and the contract was thereby made between the employer and the contractor. If the contractor then breaches the contract, it is entirely a matter between the employer and the contractor, or more likely their solicitors, to sort out. The architect has no duty other than to report to the employer as soon as there are any signs that the contractor is ceasing to work regularly and diligently and to advise the employer of his or her rights to terminate the contractor's employment or to accept the contractor's actions as repudiation. Many, possibly most, architects would throw themselves into the problem and strive to bring about an amicable conclusion. Laudable though that may be, by getting involved beyond the extent of their duties, architects run a very real danger of being blamed if their efforts are not successful. The moral is not to assume a greater responsibility than already undertaken by the terms of engagement.

184 How can an employer get rid of a contractor who seems incapable of producing good quality work?

This is a problem which seems to crop up with increasing frequency. Typically, the contractor will be unable to reach the specified quality. The architect will condemn the work again and again and eventually it is probable that something less than what is specified will be accepted out of sheer frustration. Most standard form contracts contain adequate means to enable the architect to control the quality of the work. For example, SBC in clause 3.18 gives the architect considerable powers, to order the remove from site of work not in accordance with the contract, to allow defective work to stay with an appropriate deduction to the contract sum, to order variations and to instruct opening up of other portions of the Works in order to check that there are no similar instances of defective work elsewhere. Ultimately, the architect may issue a default notice and the employer may terminate the contractor's employment if it consistently fails to rectify work. The problem is that there is nothing in any of the standard form contracts which allows the architect to form a view that the contractor is incapable of producing good quality work and to simply banish that contractor from the site. The contract assumes that the contractor is capable of good work. The clauses are designed to ensure that the architect can oblige the contractor to produce that good work and not some inferior work.

There is an assumption that the architect has carefully checked the capabilities of the contractor before inviting it to tender, requesting references from other architects as necessary. Often, the reality is that the employer will insist on a particular contractor being included in the list, because the contractor has lobbied for inclusion and indicated that it is capable of producing a low tender figure. The architect will no doubt have advised the employer that such a contractor, about which little is known, will very likely cause the expenditure of more money than is saved by the low tender. Some architects have been known to terminate their engagements at that point on the perfectly sensible view that they cannot administer a contract where the contractor is appointed against their advice and, therefore, their work is likely to be much more difficult.

The way to deal with such contractors is in the contract, as mentioned above. What tends to be missing is the correct strategy to adopt. In every contract, as soon as it becomes clear that the contractor is having difficulty achieving the specified standards, the contractor should be instructed to remove the defective work from site. This must be done in each instance of defective work. Very soon, one of two things will happen. Either the contractor will see that there is not the slightest point in trying to get away with slipshod work and it will take serious steps to improve its performance. Alternatively, a contractor who is incapable of producing decent work will start to argue that the architect is being unreasonable and looking for a standard which is in excess of what is in the contract. Such a contractor may well appeal to the employer and it is surprising how often an appeal by what the employer takes to be an experienced contractor is successful when judged against what the contractor terms the architect's 'airy-fairy ideas'. If the employer ever starts to accept the contractor's views against the architect, it is time for the architect to consider terminating his or her engagement. It is essential strictly to adhere to the strategy set out above right from the beginning of the project on site. If it is not practised until it becomes crystal clear that what appeared to be some slips in standard by the contractor are actually its best work, the contractor may be able to argue with some justification that if it has done some work of a certain standard without comment, it is unreasonable of the architect to suddenly impose a higher standard. The old adage 'start as you intend to continue' has much to recommend it.

185 Is it true that, under SBC, if the employer fails to pay, the contractor can simply walk off site?

It is surprising how often a contractor will simply stop work because it is not being paid. Generally, it is small contractors, perhaps operating under MW, who are most prone to this kind of action, but it has certainly been done with much larger contracts. It is difficult not to have considerable sympathy with a contractor who says, quite reasonably, that there is no point in doing further work if it has not been paid for what has already been done. Nevertheless, the law does not generally allow the contractor simply to stop work if not paid.

Under SBC, the remedy available to the contractor if the employer fails to pay a certificate is to issue a 7-day written notice under clause 4.14 and then to suspend performance of all its obligations if payment is not made. Additionally or alternatively, the contractor may simply issue a default notice prior to termination of its employment under clause 8.1.1. In any event, the contractor is entitled to recover interest, usually at 8 per cent above Bank of England base rate.

What is often forgotten by an employer is that the contract does not remove the contractor's common law rights. Quite the reverse. Clause 8.3.1 expressly preserves both the employer's and the contractor's other rights and remedies. It is established that if payment is made so irregularly and inadequately that the day arrives when the contractor has no confidence that it will ever be paid again, it amounts to a repudiatory breach of contract which the contractor may accept, bring its obligations to an end and recover damages.[254]

More recently, in *CJ Elvin Building Services Ltd v Peter and Alexa Noble*[255] it was held that the employer was in breach of contract, not only by refusing to pay sums as they became due, but also by threatening to make no further payment until the job was completed. The contractor was, therefore, justified in suspending the Works. Indeed, from the judgment, it was clear that the judge considered that the contractor could have accepted the employer's breach as repudiation.

254 *DR Bradley (Cable Jointing) Ltd v Jefco Mechancial Services* (1989) 6-CLD-07-19.

255 [2003] EWHC 837 (TCC).

Therefore, the next time a contractor walks off site for lack of payment, both architect and employer should think carefully about whether the contractor is simply exercising its common law rights.

186 What does 'repudiation' of a contract mean?

Repudiation is when one party to a contract makes clear by words or actions that it no longer wishes to be bound by the contract. It is a serious breach of contract, sometimes referred to as a breach of a condition of the contract or breach of a fundamental term. A typical example of repudiation would be if a contractor simply walked off site and refused to return or if a client prevented the architect from entering the site. However, repudiation is of no effect until accepted by the other party.

When faced with a repudiatory act, the innocent party has two options. It can simply affirm the contract (in other words it can carry on with the contract) and claim damages for the breach. Sometimes affirmation is not an option, as when the contractor simply walks off site and refuses to return. Alternatively, it can accept the repudiation and claim damages. When a repudiation is accepted, it puts an end to the obligations of the parties under the contract. It does not bring the contract to an end.[256]

If the other party intends to accept the repudiation, it must do so promptly and it must be careful not to do anything which can be interpreted as an affirmation of the contract. However, a party is entitled to seek legal advice before deciding whether to accept a repudiation even though that may take a little time. In some cases it is quite difficult to decide whether repudiation has taken place, particularly where the action may be open to different interpretations. For example, failure to pay an invoice cannot be repudiation if the sum is not due. If the party which submitted the invoice then purports to accept the alleged repudiation and stops work, it will itself be in repudiatory breach.[257] Delay on the part of a contractor will only be repudiatory if time is of the essence. Since time is rarely if ever of the essence in construction contracts, an employer is likely to be in repudiatory breach if purporting to accept a contractor's delay as a repudiation.[258]

256 *Photo Production Ltd v Securicor Transport Ltd* [1980] 1 All ER 556.

257 *CFW Architects (a firm) v Cowlin Construction Ltd* (2006) 105 Con LR 116.

258 *Martin John Hayes and Linda Hayes t/a Orchard Construction v Peter Gallant* [2008] EWHC 2726 (TCC).

187 Can notice of termination be sent by fax or e-mail?

Under the SBC, IC, ICD, MW and MWD forms of contract with Revision 2 there are very clear provisions for the giving of notices. Clause 8.2.3 of SBC requires every notice referred to in the termination provisions to be given in accordance with clause 1.7.4. This provides that notices must be given by hand, by Recorded Signed for or by Special Delivery post. IC, ICD, MW and MWD have provisions in similar terms. These provisions seem to preclude the use of fax or e-mail.

Before Revision 2 was introduced in 2009, the requirement was for service of such notices by actual, special or recorded delivery. Where an JCT contract earlier than Revision 2, 2009 is concerned, or indeed any other standard form which refers to 'actual delivery', it seems that delivery by fax or by e-mail is acceptable. Under IFC 98, a default notice prior to termination was given by fax and held to be validly served:[259]

> A fax, it seems to me, clearly is in writing; it produces, when it is printed out on the recipient's machine, a document, and that seems to me is clearly a notice in writing. The question is, is that actual delivery? It seems to me, if it has actually been received, it has been delivered. Delivery simply means transmission by an appropriate means so that it is received, and the evidence in this case is that the fax has actually been received. There is no dispute as to that.

Another court held that a notice of arbitration which was served by e-mail was also validly served. An important consideration in that case was that the notice was sent to an e-mail address which the recipient had indicated was its only e-mail address.[260] In that case, there was no dispute that the e-mail constituted 'writing' and the judge made no comment about that as he might have been expected to do if he had disagreed. On that basis, there seems no reason why a notice should not be validly served by e-mail provided that the contract concerned does not expressly require the notice to be served in some other form. It is not known whether either of these cases influenced the JCT to re-word its provision.

259 *Construction Partnership UK Ltd v Leek Developments* [2006] EWHC B8 (TCC).

260 *Bernuth Lines Ltd v High Seas Shipping Ltd* [2006] 1 Lloyd's Rep 537.

Chapter 24

Disputes

188 What is a dispute or difference under the contract?

SBC clause 9.2 states that if a 'dispute or difference' arises under the contract and either party wishes to refer it to adjudication, the Scheme for Construction Contracts (England and Wales) Regulations 1998 (the Scheme) will apply. Other JCT contracts have similar clauses as do other standard form construction contracts. The Scheme in paragraph 1(1) states that any party to a construction contract may give written notice of intention to refer a 'dispute' to adjudication. Therefore, it is of the greatest importance that an identifiable dispute has arisen before it can be referred to adjudication otherwise the adjudicator will not have the jurisdiction to deal with it. It might be thought that the existence or otherwise of a dispute would be fairly obvious, but questions have arisen often based on the argument that the reference to adjudication has been premature, before the dispute has crystallised. For example, a contractor may lodge a claim with an architect stating that it expects a reply within 2 weeks and, having received no response within the allotted time, issues a notice of adjudication citing the architect's failure to decide the subject matter of the claim. The first question to be considered is whether there was a dispute when the notice was issued. Was it reasonable for the architect to respond in 2 weeks? If it was not reasonable, there was no dispute and the adjudicator has no jurisdiction to decide the matter. Needless to say, this kind of question has been the subject of very many actions through the courts. Fortunately, after reviewing the legal authorities the courts have formulated a series of propositions to assist in deciding whether or not there is a dispute:

1 The word 'dispute' which occurs in many arbitration clauses and also in s. 108 of the Housing Act should be given its normal meaning. It does not have some special or unusual meaning conferred upon it by lawyers.
2 Despite the simple meaning of the word 'dispute', there has been much litigation over the years as to whether or not disputes existed in particular situations. This litigation has not generated any hard-edged legal rules as to what is or is not a dispute. However, the accumulating judicial decisions have produced helpful guidance.
3 The mere fact that one party (whom I shall call 'the claimant') notifies the other party (whom I shall call 'the respondent') of the claim does not automatically and immediately give rise to a dispute. It is clear, both as a matter of language and from judicial decisions, that a dispute does not arise unless and until it emerges that the claim is not admitted.
4 The circumstances from which it may emerge so that a claim is not admitted are Protean. For example, there may be an expressed rejection of the claim. There may be discussions between the parties from which objectively it is to be inferred that the claim is not admitted. The respondent may prevaricate, thus giving rise to the inference that he does not admit the claim. The respondent may simply remain silent for a period of time, thus giving rise to the same inference.
5 The period of time for which a respondent may remain silent before a dispute is to be inferred depends heavily upon the facts of the case and the contractual structure. Where the gist of the claim is well known and it is obviously controversial, a very short period of silence may suffice to give rise to this inference. Where the claim is notified to some agent of the respondent who has a legal duty to consider the claim independently and then give a considered response, a longer period of time may be required before it can be inferred that mere silence gives rise to a dispute.
6 If the claimant imposes upon the respondent a deadline for responding to the claim, that deadline does not have the automatic effect of curtailing what would otherwise be a reasonable time for responding. On the other hand, a stated deadline and the reason for its imposing may be relevant factors when the court comes to consider what is a reasonable time for responding.

7 If the claim as presented by the claimant is so nebulous and ill-defined that the respondent cannot sensibly respond to it, neither silence by the respondent nor even an expressed non-admission, it is likely to give rise to a dispute for the purposes of arbitration or adjudication.[261]

These propositions were approved by the Court of Appeal in *Collins (Contractors) Ltd v Baltic Quay Management (1994) Ltd*.[262] Many courts have observed that in most cases it will be obvious when there is a dispute and that the requirement that there must be a dispute will not be interpreted with legalistic rigidity. It has also been said that when the phrase 'dispute or difference' is used, it is less hard-edged than using the word 'dispute' alone.[263]

189 Is it acceptable to suggest to the nominating body whom to nominate as adjudicator?

When a party decides to seek adjudication of a dispute under a construction contract, the first stage is to send a notice of adjudication to the other party and the next stage, or sometimes concurrently, is to appoint an adjudicator. The adjudicator may be named in the contract or the contract may state the name of the body which is to nominate the adjudicator. Under the Scheme for Construction Contracts (England and Wales) Regulations 1998, if no nominating body is chosen, the referring party may chose any nominating body. Most nominating bodies have their own form which must be filled in and submitted together with a fee, after which the body chooses an adjudicator from the ones on their list and nominates that person. Often, a referring party will notify the nominating body of persons they do not wish to see nominated and it is rare that a nominating body will ignore such objections – it is easier to nominate someone to whom neither party has raised early objections. Sometimes, a referring party will ask the nominating body to nominate an adjudicator with particular qualities such as legal expertise or experience in costing or design or one who has been engaged on an earlier adjudication on the

261 *Amec Civil Engineering v Secretary of State for Transport* [2004] EWHC 2339 (TCC).
262 (2004) 99 Con LR 1 (CA).
263 *Amec Civil Engineering Ltd v Secretary of State for Transport* (2005) 101 Con LR 26 (CA).

same project. What is unusual is for a party to suggest the name of an adjudicator to the nominating body.

In *Makers UK Ltd v London Borough of Camden*[264] solicitors for the referring party decided that the adjudicator should be legally qualified. They discovered that someone they considered to be suitable was on the panel of adjudicators of the RIBA and, having ascertained that he was available, they suggested to the RIBA that he should be nominated. The RIBA did so and in due course the adjudicator accepted the appointment and proceeded to make a decision. Camden argued that the decision was void, because the adjudicator had not been properly appointed. Camden contended that there was an implied term to the effect that neither party should try to unilaterally influence the nominating body when choosing the adjudicator. This contention was rejected by the court which held that the RIBA was well able to decide whether to take notice of representations and there may well be times when representations were helpful.

Therefore, it seems that although there is nothing to stop a referring party from suggesting the name of an adjudicator, it will be relatively rare to do so.

190 An adjudicator has been appointed whom the employer has not agreed. What can the employer do about it?

The adjudicator can be appointed in various ways. The JCT Standard Building Contract 2005 (SBC) has provision in clause 9.2.1 and in the contract particulars for the adjudicator to be named, i.e. the parties have decided on the name before entering into the contract. There is also provision for the parties to simply apply to one of several bodies that maintain panels of adjudicators and which will undertake (for a modest fee) to nominate an adjudicator who has passed the criteria set by that particular body. The RIBA, RICS and the CIArb are typical nominating bodies. Of course, there is nothing to stop the parties simply agreeing the name of an adjudicator when the dispute arises. If the parties can agree a name, that is by far the best way, because the adjudicator is then a person whom both parties trust to make the right decision.

Putting the name in the contract in advance appears to have the same effect, but it does have some serious disadvantages. An

264 (2008) 120 Con LR 161.

adjudicator chosen in advance, perhaps several months or even years before required, may not be available when required due to holidays, illness, retirement or even death. The adjudicator may be admirable in very many respects, but not suited to the particular problem that arises. For example, the parties may choose Mr X, who is an architect, but the eventual dispute may concern structural steel or electrical services. The danger of simply applying to a nominating body is that the parties are stuck with whoever is nominated, unless they both object and agree to ask the adjudicator to step down in favour of another of their choosing.

Therefore, the answer to this question depends on what the contract says or, if there is no standard contract, what the Scheme for Construction Contracts (England and Wales) Regulations 1998 says.

It is essential that the adjudicator is appointed in strict conformity with any procedure that is laid down. If the relevant procedure has been complied with, there is nothing the employer can do about it. The adjudicator must be accepted. Obviously, if there is a good reason to object, for example if the adjudicator has a link with the other party, the appointment can be challenged, and if the adjudicator does not step down the employer can agree to proceed with the adjudicator without prejudice to the contention that the adjudicator lacks jurisdiction on the grounds of actual or apparent bias or an apparent breach of the rules of natural justice. Then that position can be tested in the courts if the adjudicator's decision is adverse to the employer.

If the proper procedure for the appointment of the adjudicator has not been observed, the employer is on firm ground to challenge and the adjudicator should resign as soon as it is established that the appointment is flawed.

191 How important are the various time periods in adjudication?

The adjudication process is set to a very strict and tight timetable. The process is started by the serving of a notice of adjudication and the adjudicator must be appointed and the referral served on the adjudicator within 7 days. The adjudicator must reach a decision within 28 days of the date of the referral, but that may be extended by 14 days if the referring party consents or to any date provided that it is agreed by both referring party and responding party. These are basic time periods under the Housing Grants, Construction and Regeneration Act 1996, but the Scheme for Construction Contracts (England and Wales) Regulations 1998 and the provisions of various procedures and standard form construction contracts may add additional times to those in the Act. For example, the JCT Standard Form of Contract 1998 stipulated that the responding party had 7 days from receipt of the referral in which to respond. In a case under another contract with the same provision, the court held that the adjudicator had absolute discretion to extend the period for service of the response, because the adjudicator had the right under the contract to 'set a procedure'.[265] Part 8 of the Local Democracy, Economic Development and Construction Act 2009, which at the time of writing had not commenced, does not amend any of the statutory time periods for adjudication.

The general rule is that where time periods are specified, they are very important and failure to observe them may render the adjudication process, or any decision, a nullity. For example, if the referral is served later than the 7th day after the notice of adjudication, it will be invalid and the adjudicator will have no jurisdiction.[266] However, that quite stark position was modified by a court when considering the situation where the referral notice had been served on time to adjudicator and responding party, but where the adjudicator had not received the accompanying documents until 3 days later.[267] The referral was held to be validly served. An important factor was that the responding party had received the referral and the accompanying documents on time.

The adjudicator must reach a decision 28 days after the date of the referral notice. It is important to note that this period starts from the date of the notice, not from the date the notice is received by the adjudicator. Paragraph 19(3) of the Scheme requires the adjudicator to deliver a copy of the decision to the parties as soon as possible after the decision has been reached. There has been considerable discussion about the position if the adjudicator fails to reach a decision in time or if the decision is reached in time but is not delivered as soon as possible. The problem has been the subject of several cases. It now appears settled that the adjudicator's jurisdiction to make a decision comes to an end on the expiry of the time

265 *CJP Builders Ltd v William Verry Ltd* [2008] EWHC 2025 (TCC).
266 *Hart Investments Ltd v Fidler and Another* (2006) 109 Con LR 67.
267 *Christopher Michael Linnett v Halliwells LLP* (2009) 123 Con LR 104.

limit if it is not already extended.[268] Attempts by a party to extend the adjudicator's time will be ineffective if the period has already expired, because something which no longer exists cannot be extended. So far as delivery is concerned, if the adjudicator reached a decision within the relevant timescale, a 2-day delay in delivering the decision to the parties was not sufficient to render the decision a nullity.[269] The moral is to observe the time periods precisely as set out in the particular contract between the parties. If the contract specifies that a particular procedure should be adopted, such as the Scheme, then the time periods in the Scheme should be followed. Just because a court has allowed a day or two grace in a particular circumstance is no guarantee that a subsequent court will not decide that the circumstances it has to consider are markedly different.

192 What exactly is a failure to observe the rules of natural justice?

The phrase is now usually associated with the work of an adjudicator although the concept applies equally to arbitrators who are obliged to act fairly and impartially giving each side a reasonable opportunity to put their case and deal with that of the other side. Natural justice is an obligation, upon those with power to take decisions which affect others, to act with procedural fairness. A failure to act in accordance with natural justice is, like the elephant, easier to recognise than to describe. Most people have an instinctive feel for natural justice and can readily see when it is being ignored. So far as adjudicators and arbitrators working in the construction industry are concerned natural justice includes:

- acting impartially, without bias and in good faith;
- conducting proceedings fairly, giving both parties the opportunity to present their cases and sufficiently comment on the contentions and evidence presented by the other side.

268 *Ritchie Brothers PWC Ltd v David Philp (Commercials) Ltd* [2005] BLR 384.

269 *St Andrews Bay Development Ltd v HBG Management Ltd and Miss Janey Milligan* [2003] Scot CS 103; *Barnes & Elliot Ltd v Taylor Woodrow Holdings Ltd* [2004] BLR 111.

An adjudicator is obliged to act impartially.[270] However, adjudicators are obliged to come to a decision in 28 days from the date of the referral unless the time is extended. Even where an extension of the 28 days is agreed, an adjudicator will not have a great deal of time to consider the submissions and possibly investigate the issues. Working to a tight timetable is the norm. Therefore, the courts tend to interpret the duty of an adjudicator to comply with the rules of natural justice flexibly. The courts will rarely interfere with an adjudicator's decision. But they will not enforce a decision if an adjudicator has breached natural justice to any significant degree.[271]

Telephone calls between one party and the adjudicator should be avoided, because they are easily construed as indicating a bias on the part of the adjudicator. Most adjudicators make clear when accepting the appointment that they will not accept telephone calls from either party unless such calls are initiated by the adjudicator as conference calls. Where a party is inexperienced in the niceties of adjudication, they might telephone the adjudicator seeking advice about procedure or even arguing their case. An adjudicator receiving such a call must, of course, decline to discuss the matter and terminate the call politely but firmly followed by a letter to both parties confirming the call, the length and what was said. Sometimes an adjudicator may believe that it is quicker and easier to telephone one party with a request than to write to both. This might be the case where one party has neither e-mail nor fax facilities and writing would take too long. If it is necessary for such calls to take place there is no reason why they should cause substantial injustice and will not usually constitute a breach of natural justice, but they must be confirmed to both parties as noted above.

It is rare that an adjudicator will fail to give both parties equal opportunity to put their cases, but it may happen, particularly if there have been numerous exchanges and the adjudicator refuses to consider a late submission. A failure to take into account all the evidence will not necessarily preclude enforcement of a decision if that

270 S 108(2)(e) of the Housing Grants, Construction and Regeneration Act 1996 and clause 12 of the Scheme for Construction Contracts (England and Wales) Regulations 1998 (and the Northern Ireland and Scottish equivalents). The Local Democracy, Economic Development and Construction Act 2009 does not change that position.

271 *Discain Project Services Ltd v Opecprime Developments Ltd (No. 2)* [2001] BLR 285.

results from an error of law.[272] It is a brave adjudicator who refuses to consider any submission unless there is an extremely good reason. Such a reason might be where both parties have had equal chance to put their positions and the adjudicator has notified them that no further submissions are required and one of the parties submits a lengthy further submission just before the adjudicator has to reach a decision.

Some adjudicators have seen their decisions invalidated, because they have decided to look into the dispute and performed calculations without reference to the parties.[273] Where an adjudicator believes that there are issues which the parties have not properly addressed in submissions, these issues should be referred back to the parties together with any new matters the adjudicator thinks might be relevant. For example, adjudicators commonly come across a legal case which appears relevant, but which neither party has mentioned. In such instances, the adjudicator should notify the parties and seek their views before making the decision.

193 Is the architect obliged to respond to the referral on behalf of the employer if so requested?

This is not as unusual a question as may seem at first sight. Most employers have no idea about adjudication. When they receive a notice of intention to seek adjudication from a contractor, they immediately send it to the architect who has been handling the project for them. No doubt, if there is a project manager, the employer will direct the notice there instead. But, whether project manager or architect, is there a duty to deal with the adjudication? Clearly the answer is 'No'.

So far as the architect is concerned, his or her duties will probably include taking the brief, preparing designs, submitting for planning and building control, preparing detailed construction drawings and administering the contract with all that entails, including dealing with contractor's claims (usually at an additional fee). It is understood that the architect is skilled and experienced at doing all these things. But the architect has neither the skills nor the training to run

272 *Kier Regional Ltd (t/a Wallis) v City and General (Holborn) Ltd* [2006] BLR 315.

273 *Primus Build Ltd v Pompey Centre Ltd & Slidesilver Ltd* [2009] EWHC 1487 (TCC).

an adjudication. The architect is no more capable of doing that than running an arbitration or dealing with court procedures. None of these things will be among the duties the architect has agreed to carry out. Importantly, the architect will have no professional indemnity insurance to cover such work.

Some architects may have special expertise in dealing with adjudications, just as some architects have expertise at acting as expert witnesses or designing particular types of buildings. In that case, there is no reason why such architects should not run the adjudication on behalf of the employer (again, for an additional fee). They should always remember that the adjudication is between the contractor and the employer, not between the contractor and the architect, even though the contractor may be invoking adjudication to question an architect's decision.

Although architects have no duty to run the adjudication for the employer, they clearly do have a duty to give the employer basic initial advice when the notice is brought to their attention. Architects should be capable of doing that, and it will usually consist of advising their clients to seek appropriate legal advice without delay in view of the alarmingly short timescale involved.

Architects, together with all the other consultants, certainly have duties to assist the employer's legal advisers by providing copy correspondence, drawings and, if appropriate, explanations about various aspects of the project. An appropriate fee will be chargeable, probably on a time basis. However, where the information required by the employer's legal advisers is more than merely factual (correspondence and drawings) consultants should tread warily. If the adjudicator's decision goes against the employer, the employer's legal advisers may start to look very carefully at the opinions and reports provided by consultants to see if the employer has a means of redress in that direction. It is not going too far to suggest that an architect or other consultant who is asked to assist in this way should consult their own advisers with a view to protecting their positions in the future.

An architect recently asked whether or not there was a duty to assist the employer in an adjudication that had been commenced by the contractor some time after the employer had decided to do without the services of the architect half-way through the project. The employer's legal advisers were pressing the architect to attend meetings and even to decide matters of extension of time and loss and/or expense. In such situations, the architect has no duty other than providing the employer with copies of documents on payment of reasonable copying expenses. By dispensing with the architect's services, the employer was no longer relying on the architect's skill and care. If the architect was asked to provide a witness statement as to facts, it is arguable that the architect could refuse. An architect asked to give evidence in arbitration or litigation could face a subpoena to do so. But an employer who had sacked the architect would probably have little to gain by trying to force the architect to give evidence in such cases.

194 An Adjudicator's Decision has just been received and it is clear that the points made have been misunderstood and the adjudicator has got the facts wrong. Can enforcement be resisted?

The short answer to this is 'No', at least not on those grounds. When an adjudicator makes a decision, the parties must carry out that decision within whatever timescale the adjudicator has laid down. If a party fails to carry out the decision, the other party has the right to apply to the court to enforce the decision. The decision of an adjudicator, while not quite sacrosanct, is at least protected until the parties decide that one or other wants to have the dispute finally decided in arbitration or in legal proceedings.

Adjudicators are human like the rest of us. Sadly, some adjudicators seem to have a tenuous grasp of the law. Nonetheless, when an adjudicator has been nominated, or perhaps the parties have agreed the name between them, that person is the one entrusted to make a decision on the merits of the dispute. It may well be that the adjudicator misunderstands some of the points made, and some participants do not make themselves very clear. On the other hand, some participants are represented by experts who subject the adjudicator to a barrage of words.

It is not unusual for an adjudicator to receive half a dozen or more lever arch files as the referral and a couple of similar-sized files as the response, to say nothing of a multitude of submissions on jurisdiction. It is little wonder if some of the more subtle points are overlooked in this scenario. Again, it must be said that some adjudicators are not good at handling clever points and generally try to come to a decision on what they believe is the overall justice of the case. Of course, this is not what an adjudicator is supposed to do. An adjudicator, just like an arbitrator or a judge, is charged with applying the law, not a personal gut feeling. Lord Denning may have been famous for that very thing, but he was an exception and in any event he could never have been accused of not knowing the law.

Adjudication was never intended for this kind of detailed argument. It was originally devised as a method of getting a quick result for problems that regularly bedevil the construction industry. The principle is that, if the adjudicator answers the right question in the wrong way, the decision will be upheld by the courts, but if the adjudicator answers the wrong question in the correct way, the decision will be a nullity. Another way to put it is that the adjudicator can answer only the question asked in the notice of intention to seek adjudication. Neither party can unilaterally introduce new questions; there can be no counterclaims. The only exception is if the adjudicator has to answer a question that has not been asked in order to answer the question asked.[274] Even if the adjudicator has misunderstood the facts or simply got them wrong, it is not enough to resist enforcement of what might well be a flawed decision.[275]

Basically, it is a policy decision. Adjudication is a quick method of settling disputes, but a comparatively coarse remedy. With this kind of remedy, the parties have to accept that there will be rough justice and, occasionally, even bad errors. If the money at stake is sufficiently large, no doubt the parties will seek a solution in a forum that can give proper time and consideration to the arguments.

The main acceptable grounds for resisting enforcement of a decision tend to concern whether the adjudicator had the jurisdiction to make the decision. If it can be shown that there was no jurisdiction, the enforcement can be resisted, because there is no decision. Lack of jurisdiction is often due to a failure to answer the question asked, so that the adjudicator has no jurisdiction for the decision about an unasked question. It may also be because it can be shown that there was no dispute in being at the time the notice of intention to seek adjudication was issued. Successful challenges have also been made on the basis that the adjudicator was in breach of the rules of natural

274 *Karl Construction (Scotland) Ltd v Sweeney Civil Engineering (Scotland) Ltd* (2000 and 2002) [CS (OH and IH)] 85 Con LR 59.

275 *Bouygues (UK) Ltd v Dahl-Jensen (UK) Ltd (in Liquidation)* (2000) 73 Con LR 135.

justice. This can be quite complicated in practice, but in essence it refers to the need for each party to have an opportunity to put its case and the fact that the adjudicator must not discuss the dispute with one party in the absence of the other. Obviously, a decision will also be thrown out if it can be shown that an adjudicator was biased in favour of the other party.

195 Can an adjudicator make a decision about an interim payment if it is the final account value which is being referred?

It is well understood that an adjudicator only has the power to consider the question which has been referred. That is the dispute identified in the notice of adjudication. However, if it is necessary to decide other things as steps in reaching the answer to the dispute referred, the adjudicator is entitled to consider these other things. There is an important difference between the name of a dispute and what the dispute actually is. For example, it is common for contractors and others to refer to the 'penalty clause' when what they actually mean is the 'liquidated damages clause'. The courts understand this and regularly have to analyse a situation to decide what the parties actually mean.

The straight answer to this question is that an adjudicator cannot decide a question about an interim payment if it is the final account which has been referred. However, the final account and the last interim payment are often confused. This seems to have occurred in a recent case where the referring party, in its notice of adjudication, stated that the dispute related to the final account and then requested the adjudicator to order a specific payment or such sum as the adjudicator may decide.[276] The adjudicator stated that the decision was given in respect of an interim application based on the final account. The court had no difficulty in deciding that the adjudicator had decided what was referred whether it was called the final account or the interim payment based on the current draft of the final account. Each question will depend on particular circumstances, but it is clear that the courts will take a commonsense approach.

276 *OSC Building Services Ltd v Interior Dimensions Contracts Ltd* [2009] EWHC 248 (TCC).

196 Other than going to arbitration, are the parties stuck with an adjudication decision which contains obvious errors in calculations?

Yes, unless the adjudicator agrees to amend the decision in accordance with the 'slip rule'. If one of the parties draws the adjudicator's attention to a clear arithmetical or similar kind of error within a day or two of the decision being issued, or if the adjudicator spots the error and promptly corrects it, the correction will be valid.[277] The idea is that the adjudicator can correct the decision in order to give effect to his first thoughts. Such things as accidental transposition of names or failing to give credit for sums already paid would be included in the sort of slips which an adjudicator can correct if acting promptly. Section 140 of the Local Democracy, Economic Development and Construction Act 2009, which was not in force at the time of writing, amends the adjudication provisions of the Housing Grants, Construction and Regeneration Act 1996 to expressly allow the adjudicator to correct clerical or typographical errors arising by accident or omission. However, an adjudicator cannot use the slip rule to correct a decision simply because of second thoughts. After making the decision, the adjudicator may read a case which makes clear that the decision is wrong in law. There is nothing that the adjudicator can do in such circumstances. Even though an adjudicator may reach a decision which is clearly wrong, a court will not interfere with it provided it is a decision which the adjudicator had the jurisdiction to make. If the adjudicator answers the wrong question, even though giving the right answer, the decision will be a nullity, but if the adjudicator gives the wrong answer to the right question, it will be enforced.[278]

197 Can the losing party set off monies owing against the adjudicator's order requiring payment?

The point of an adjudicator's decision is that it is quick justice; one might say rough justice at times. However, it seems that most participants accept the adjudicator's decision and although some challenge

277 *YCMS Ltd (t/a Young Construction Management Services) v Grabiner and Another* (2009) 123 Con LR 202.

278 *Bouygues (UK) Ltd v Dahl-Jensen (UK) Ltd* (2000) 73 Con LR 135.

the decision in the courts on the grounds of lack of jurisdiction or breach of the rules of natural justice, few decide to refer the dispute anew to arbitration or legal proceedings through the courts. Various means have been tried to avoid payment, but the courts have been reluctant to allow such attempts to succeed. The decision of the Court of Appeal in *Ferson Contractors Ltd v Levolux AT Ltd*[279] sets the tone:

> The contract must be construed so as to give effect to the intention of Parliament rather than to defeat it. If that cannot be achieved by way of construction, then the offending clause must be struck down. I would suggest that it can be done without the need to strike out any particular clause . . .

That case was dealing with an attempt to avoid payment by appealing to a clause in the contract which appeared to preclude further payments. It is clear that the Court of Appeal was prepared simply to effectively ignore such a clause. From time to time, a party will try to avoid payment by trying to set off from a payment ordered by the adjudicator a payment it believes is due to it. Many cases have been heard by the courts on this basis and the attitude of the courts has been to refuse to allow set-off which would interfere with the adjudicator's decision. In a recent case, the court ruled that set-off of liquidated damages against the adjudicator's decision that a sum was due under the contract was inadmissible. Clause 41A.7.2 of the JCT 98 contract expressly stated that the parties must comply with the adjudicator's decision and there was no reference to any right of set-off. A notice of set-off had been served, but it had not been given 7 days before the final date for payment (in accordance with the Scheme) stipulated by the adjudicator. The notice had in fact been issued only 6 days before the final date. The fact that it was impossible to issue the notice earlier because the adjudicator had allowed only 7 days for payment cut no ice with the court.[280] There may be instances where a court will be persuaded to allow a set-off against the adjudicator's decision, but such instances will be rare.

279 (2003) 19 Const LJ T83.

280 *Avoncroft Construction Ltd v Sharba Homes (CN) Ltd* (2008) 119 Con LR 130.

198 If the adjudicator is late with the decision, is it still valid?

Under section 108 of the Housing Grants, Construction and Regeneration Act 1996, the adjudicator must reach a decision within 28 days of the date of the referral. This is subject to the right of the referring party to agree an extension this time period for up to a further 14 days or for both parties to agree an extension of the period of whatever length as they decide. The Scheme for Construction Contracts (England and Wales) Regulations 1998, at paragraph 19(3), provides that the adjudicator must deliver a copy of the decision to each party 'as soon as possible' after reaching it. The courts have come to different conclusions about the position if the adjudicator fails to reach the decision within the allotted time or, having reached the decision on time, fails to communicate it to the parties on time. It has been suggested that the decision of an adjudicator will be valid provided that it is reached before the parties have terminated the adjudication agreement. That particular view is not thought to be a correct statement of the law. Indeed, the correct position and probably the safest view to take for adjudicators and parties alike is that the words in the Act mean what they say and that the 28-day period is mandatory in that the decision must be reached within the 28-day period or within any validly extended period. Decisions reached after the expiry of the period are almost certainly void, because the adjudicator's jurisdiction comes to an end at the end of the period whether or not a decision has been made.[281] However, it seems that the delivery of the decision to the parties may take place after the 28-day period provided that it is delivered without any undue delay. In practice, that means that it must be delivered with all practicable speed. There is little excuse for late delivery of a decision when virtually instantaneous transmission is possible by fax or e-mail.

199 The court has just ruled that the Adjudicator's Decision is a nullity; can the losing party refuse to pay the adjudicator's fees?

This is an interesting question and the answer is not yet entirely clear. Adjudicators sometimes have trouble in recovering their fees,

281 *Epping Electrical Company Ltd v Briggs & Forrester (Plumbing Services) Ltd* (2007) 113 Con LR 1.

because they are not entitled to withhold the decision pending receipt of the fee as an arbitrator is entitled to do. On the other hand, the parties to an adjudication are jointly and severally liable for the fee, therefore, if the adjudicator fails to recover it from the losing party, it can be recovered from the winner. The position if an adjudicator's decision was found to be a nullity was briefly considered in *Dr Peter Rankilor & Perco Engineering Service Ltd v M Igoe Ltd.*[282] In that case, the court held that the decision was valid, but commented:

> I do not need to consider Dr Rankilor's claim that even if his decision had been vitiated by a breach of the rules of natural justice, he would still have been entitled to recover his fee under the terms of the adjudication contract which had been entered into. It is, I must say, a surprising submission that if an adjudicator's decision has been reached in serious breach of the rules of natural justice and thus would not be enforced by the court, that the adjudicator should nevertheless be entitled to claim payment for producing what was in fact a worthless decision without even any temporary binding legal effect. I prefer however to leave that question for determination in a case where it is necessary to do so. The present is not such a case.

The extent to which a party is liable for the adjudicator's fees, even where jurisdiction is challenged, has been considered in a later case and a number of guidelines have been laid down[283]:

- The right of the adjudicator to be paid depends on the contract with the parties.
- Even if a party challenged the adjudicator's jurisdiction, continued participation in the adjudication would give rise to an implied agreement to pay a reasonable sum. That would usually be because there was a contract, but it would also be justified on the basis of unjust enrichment: a responding party has been enriched by receiving a decision from the adjudicator on which it can, if it wishes, rely. The enrichment is at the expense of the adjudicator who has spent time and incurred cost dealing with the submissions. If the benefit is accepted without payment, it is unjust enrichment.

282 [2006] EWHC 947 (TCC).
283 *Christopher Michael Linnett v Halliwells LLP* (2009) 123 Con LR 104.

- Where the responding party challenges jurisdiction and takes no further part in the adjudication, the adjudicator will still have a claim for fees against the referring party on the basis that the work was carried out at its request.
- A party which subsequently relied on the adjudicator's decision would be bound by it in any event.

The court appears to have been keen to demonstrate a number of different ways in which an adjudicator can claim fees. It suggests that even if the decision is a nullity, the referring party, if not the responding party, will be liable for the fees although it leaves open whether the referring party could argue that it is not liable to pay fees to an adjudicator whose action caused the decision to be a nullity.

200 If the contractor wants to take matters beyond adjudication, what are the pros and cons of arbitration and litigation?

If the parties wish to achieve a final and binding decision, the choice is between arbitration and legal proceedings. Under the 1998 suite of JCT contracts, the default position, if no choice was made, was to arbitration. That position changed with the 2005 JCT contracts and now legal proceedings apply unless the contract particulars are completed to show that arbitration is to be the dispute resolution procedure. The decision to default to legal proceedings rather than arbitration is seen by many as a backward step, because arbitration has always been the tribunal of choice for technical matters. Employers are generally well advised to complete the contract particulars so as to refer disputes to arbitration. It should be noted that, if the parties have agreed that the method of dispute resolution will be arbitration, attempts to use legal proceedings instead may be blocked by the other party in reliance on section 9 of the Arbitration Act 1996.[284] Section 9 provides that the court must grant a stay (postponement) of legal proceedings until the arbitration is concluded unless the arbitration is null, void, inoperable or incapable of being performed. Under the 1996 Act, the court has no discretion about the matter and the successful party should recover its costs. The advantages of arbitration are usually:

284 *Ahmad Al-Naimi v Islamic Press Agency Incorporated* [2000] BLR 150.

- *The parties are in control* – The parties can decide between themselves timescales, procedure and location of any hearing.
- *Speed in reaching a decision* – This depends on the arbitrator and on the will of the parties; but a good arbitrator should deal with most cases in a matter of months, rather than years.
- *Technical expertise of the arbitrator* – The fact that the arbitrator understands construction should shorten the time schedule and may avoid the need for expert witnesses if the parties agree that the arbitrator can use his or her own knowledge.
- *Privacy* – Only the parties and the arbitrator know the details of the dispute, the arguments and the final arbitrator's award. Sensitive matters can be kept out of the public eye.
- *Expense* – In theory, arbitration should be more expensive than legal proceedings, because the parties (usually the losing party) have to pay for the arbitrator and the hire of a room. However, an efficient arbitrator with appropriate technical knowledge usually keeps the costs down.
- *Appeal* – The arbitrator's award is usually final, because the courts are reluctant to consider any appeal.[285]

Disadvantages of arbitration can be:

- *Arbitrator appointed by a third party* – Parties who are in dispute often find it difficult to agree about anything. Therefore, the arbitrator may be appointed by the appointing body and the procedure, the timing and the location of the hearing room may be decided by the arbitrator with the result that neither party is satisfied. Parties to legal proceedings cannot choose the judge.
- *Slow process* – If the arbitrator is not very good, the process may be slow and ineffective and fail to produce a satisfactory result. The answer is for the parties to choose an arbitrator of known reputation.
- *Arbitrator not necessarily an expert on the law* – That may be a crucial part of the dispute. The result may be a poor award. The answer is for the parties to choose the arbitrator with care.
- *Expensive* – The parties (usually the losing party) pay the cost of the arbitrator and the hire of a room for the hearing. An efficient arbitrator can maximise resources including the rooms.

285 *The Council of the City of Plymouth v D R Jones (Yeovil) Ltd* [2005] EWHC 2356 (TCC).

The advantages of legal proceedings are usually:

- *The judge is an expert on the law* – Therefore, there should be no worries that the judge will not understand a difficult legal point.
- *Experienced judges* – Many of the judges in the Technology and Construction Courts have a sound understanding of construction matters.
- *Civil Procedure Rules* – They govern legal proceedings and require judges to manage their caseloads and encourage pre-action settlement through use of the Pre-Action Protocol. This may end in adjudication rather than legal proceedings.
- *Speed* – It is said that cases can reach trial quickly. People who have been through the legal system are rarely convinced about this.
- *Multi-party actions* – The claimant can join several defendants into the proceedings to allow interlocking matters and defendants to be decided.
- *Expense* – Costs of the judge and courtroom are minimal.
- *Appeal* – A dissatisfied party can appeal to a higher court. This may be a mixed blessing.

The disadvantages of legal proceedings are said to be:

- *Lack of technical knowledge* – Even specialist judges know relatively little about the details of construction work.
- *No choice of judge* – Parties cannot choose the judge, who may be unsuitable for the case.
- *Slow* – Cases often take a long time to resolve.
- *Expense* – Costs will be increased, because expert witnesses or a court appointed expert witness will be required to assist the judge to understand relatively simple points and lengthy timescale and complex processes may result in high costs.
- *Appeals* – The possibility of appeals may make finality difficult or slow to achieve and result in an unacceptable level of costs.

With the advent of adjudication as the major, albeit not final, method of dispute resolution, less consideration is being given to the choice between arbitration and legal proceedings as the finally binding method. This is a topic which architects should be careful to discuss with their clients at the time procurement systems and the various forms of contract are being considered.

Table of cases

Index